Fluid Mechanics at Interfaces 1

Fluid Mechanics at Interfaces 1

Methods and Diversity

Edited by

Roger Prud'homme
Stéphane Vincent

ISTE

WILEY

First published 2022 in Great Britain and the United States by ISTE Ltd and John Wiley & Sons, Inc.

ISTE Ltd
27-37 St George's Road
London SW19 4EU
UK

www.iste.co.uk

John Wiley & Sons, Inc.
111 River Street
Hoboken, NJ 07030
USA

www.wiley.com

Library of Congress Control Number: 2021949304

British Library Cataloguing-in-Publication Data
A CIP record for this book is available from the British Library
ISBN 978-1-78630-816-0

MIX
Paper from responsible sources
FSC® C013604

Contents

Preface ix

Roger PRUD'HOMME, Stéphane VINCENT, Christian CHAUVEAU and Mahouton Norbert HOUNKONNOU

Chapter 1. Modeling Interfaces with Fluid Phase 1

Roger PRUD'HOMME

1.1. The concept of an interface 3
1.1.1. Interface in physics and geometric surfaces 3
1.1.2. The concept of equilibrium in the domain of interfaces 6
1.2. Some examples of interfaces. 6
1.2.1. Fluid phase change and separation interface 6
1.2.2. Solidification interface. 7
1.2.3. Generalized interfaces 8
1.3. Mathematical description of an interfacial layer 10
1.3.1. Normal gradient and tangential gradient 12
1.3.2. Structure and kinematics of an interfacial layer 13
1.3.3. Bulk and surface quantities 16
1.3.4. Interface balances. 17
1.3.5. Constitutive laws 20
1.4. Some additional information and examples of application 22
1.4.1. Effective surface tension between two miscible liquids 22
1.4.2. Terms that come into play in interface balance laws 23
1.4.3. Normal shockwave 24
1.4.4. Combustion waves 26
1.4.5. Thin premixed flame 29
1.4.6. Boundary layers. 31
1.5. Conclusion 31
1.6. References 33

Chapter 2. Simulations of Turbulent Two-Phase Flows with Phase Change Using a Multifield Approach Combined with LES 37

Solène GOUÉNARD, Stéphane VINCENT and Stéphane MIMOUNI

2.1. Introduction. 40
2.2. Computational model 42
2.2.1. Two-fluid model 43
2.2.2. Large Bubble Model 44
2.3. Filtered two-fluid equations 46
2.4. *A priori* LES study 47
2.4.1. Presentation of the test case. 48
2.4.2. Order of magnitude of the subgrid terms 49
2.4.3. Comparison of turbulence models 52
2.4.4. ADM order. 56
2.4.5. Effect of the filter. 58
2.5. Comparison of turbulence models with true LES 61
2.5.1. Presentation of the METERO experiment 62
2.5.2. Presentation of the test case. 63
2.5.3. Simulation results. 63
2.5.4. Comparison between RANS and LES. 64
2.5.5. ADM implementation 66
2.6. New phase change model for large interfaces. 68
2.6.1. Implementation of the new heat flux model 68
2.6.2. Validation of the new heat transfer model 70
2.6.3. Sucking problem 71
2.6.4. Stefan problem 74
2.7. Conclusion 77
2.8. References 78

Chapter 3. An Original Approach to Extract Momentum and Heat Transfers from Particle-Resolved Simulations of Particulate Flows 83

Mohamed-Amine CHADIL, Stéphane VINCENT and Jean-Luc ESTIVALÈZES

3.1. Introduction. 83
3.2. Numerical methodology 86
3.2.1. Viscous penalty method 86
3.2.2. Drag force and heat flux computation using Aslam extension . . . 88
3.3. Isolated stationary sphere passed by a uniform flow. 99
3.3.1. Drag force computation 100
3.3.2. Heat transfer computation. 105
3.4. Face-centered cubic arrangement of stationary sphere passed by a uniform flow. 107

3.4.1. Monodispersed face-centered cubic periodic arrangement of spheres . . . 108
3.4.2. Bidisperse face-centered cubic periodic arrangement of spheres . . 111
3.5. Conclusion . . . 113
3.6. Acknowledgments . . . 114
3.7. References . . . 114

Chapter 4. Interfaces and Critical Fluids . . . 121

Roger PRUD'HOMME

4.1. Thermostatics of fluids in the vicinity of the critical point . . . 123
4.1.1. Real fluids . . . 123
4.1.2. A van der Waals fluid . . . 125
4.1.3. Other laws for gases and dense liquids . . . 128
4.2. Thermodynamics of fluids in the vicinity of the critical point . . . 129
4.2.1. Universal exponents . . . 129
4.2.2. Isobaric evolutions . . . 131
4.2.3. Valid expressions of varying distance from the critical point . . . 131
4.2.4. Summary of theoretical approaches . . . 132
4.3. A specific mode of heat transmission: the piston effect . . . 132
4.4. Expansion of a "drop" at critical pressure . . . 136
4.5. Behavior of a pocket of supercritical fluid immersed into a high-temperature environment . . . 139
4.6. Boiling near the critical point . . . 142
4.7. Conclusion . . . 146
4.8. References . . . 146

Chapter 5. Shear-Induced Anomalies in the Brownian Motion of Particles in Strongly Fluctuating Near-Critical Mixtures . . . 151

Daniel BEYSENS

5.1. Introduction . . . 151
5.2. Theoretical background . . . 152
5.2.1. Miscibility critical point in binary mixtures . . . 152
5.2.2. Effect of shear flow . . . 154
5.2.3. Colloid Brownian motion and shear flow . . . 156
5.3. Experiments and methods . . . 158
5.4. Results and discussion . . . 161
5.5. Concluding remarks . . . 163
5.6. Acknowledgements . . . 163
5.7. Appendix: Light scattering (photon beating spectroscopy) . . . 163
5.8. References . . . 165

Chapter 6. Basics on Interfaces in Combustion 167

Roger PRUD'HOMME

6.1. Introduction. 168
6.2. Non-premixed laminar combustion. 172
6.2.1. Overview of a candle flame 172
6.2.2. Analytical solution of a diffusion flame: the Burke–Schumann problem 173
6.2.3. Numerical solutions 176
6.3. Turbulent non-premixed combustion. 176
6.4. Premixed combustion 179
6.4.1. Propagation of a premixed flame. 181
6.4.2. The combustion rate of the adiabatic, planar premixed flame 183
6.4.3. Concepts in turbulent premixed combustion 187
6.5. Plate combustion. 188
6.6. Powders 191
6.6.1. Thermites. 191
6.6.2. Solid propellant rockets 192
6.6.3. Spin-like combustion. 196
6.7. Sprays and fireworks 197
6.8. Conclusion 199
6.9. Acknowledgments. 200
6.10. Appendices 201
6.10.1. Appendix A: The Rankine–Hugoniot theory 201
6.10.2. Appendix B: Historical overview of research on combustion . . . 212
6.11. References. 214

List of Authors 219

Index 221

Summary of Volume 2 223

Preface

In this book, it is relevant to recall the general objectives of the Systems and Engineering e-journal *Thermodynamics of Interfaces and Fluid Mechanics*[1]. The thermodynamics of interfaces and fluid mechanics deals with interfaces, thin spaces that separate media with different properties. These interfaces can denote phase separations, as well as thin flames and discontinuity waves. On a microscopic scale, they can be compared to physical surfaces that are endowed with thermodynamic properties and possess their own constitutive laws. The analysis of systems that contain interfaces involves scale changes and the use of specific techniques, such as asymptotic expansions, the second gradient theory and phase field models. Numerical simulations are used to solve the complex systems studied. Experimentation is an essential step in solving the problems posed. The 2D manifolds that the interfaces form often coexist with 1D manifolds, such as ligaments (atomization), contact lines (drops placed in a line) or Plateau edges (foam).

The achievement of the general objectives of *Thermodynamics of Interfaces and Fluid Mechanics* is partly concretized in the first volumes of the *Fluid Mechanics at Interfaces* series of books. However, the OpenScience journal has a very broad scope and the articles, published here as chapters (reworked by the authors, from French to English, which requires the modification of certain parts of the articles and updated, etc.), represent only a small part of the interface world, the focus resulting from the authors' choices. The classification of chapters does not follow the chronology of the articles, but has been organized to be coherent.

1 To consult these articles, see: https://www.openscience.fr/Thermodynamique-des-interfaces-et-mecanique-des-fluides.

The first volume, subtitled *Methods and Diversity*, focuses more on the sciences, with their various methods of analysis, depending on the scales of space, speed and time, from the microscopic or small scale (starting with molecular and nanoscopic scales characteristic of the matter) to the macroscopic (including meso and interstellar scales), as well as their laws (classical mechanics, quantum mechanics and relativity).

The chapters are organized as follows:

Chapter 1 (R. Prud'homme) is an introduction to *Fluid Mechanics at Interfaces*, and as such, we examine questions raised by the modeling of interfaces in the presence of one or more fluid phases. The following themes are examined: scales of length and time, the definition of interfacial quantities, microscopic and macroscopic balance equations and constitutive laws. These are required to close the systems of equations that we obtain. Solutions are obtained by using asymptotic expansions and the virtual power method.

Chapter 2 (S. Gouénard, S. Vincent, S. Mimouni) studies two-phase flows. While models have already been developed to study liquid–vapor flows that contain both small, dispersed bubbles and large bubbles (the large bubble model coupled with a dispersed model), the literature has been limited, so far, to laminar regimes. The study presented here relates to the action of turbulence in these two-phase flows, taking into account heat exchanges at the interfaces, as occur in the secondary cooling circuits (exchangers) of nuclear power plants. A new model, using the large-eddy simulation method, is presented here.

Chapter 3 (M.-A. Chadil, S. Vincent, J.-L. Estivalèzes) studies an original method for calculating the drag force and thermal transfers in flows around the networks of spherical particles. This chapter also examines numerical approaches in a two-phase fluid/solid medium, using particle-resolved methods, i.e. the size of the particles is greater than that of the cells in the calculation mesh. An original approach is proposed to extract momentum and heat transfers from particle-resolved simulations of particulate flows.

Chapter 4 (R. Prud'homme) studies the relationships between interfaces and critical fluids. The vicinity of the critical liquid–vapor point of pure bodies presents unique behaviors. In particular, the interface between the liquid and its vapor vanishes when passing the critical point. The piston effect replaces the conductive transfer of heat in the vicinity of solid walls in a 1D configuration.

Chapter 5 (D. Beysens) examines shearing that causes Brownian motion anomalies in particles immersed in fluids that are undergoing strong fluctuations close to their critical point.

Chapter 6 (R. Prud'homme) introduces basic concepts related to combustion interfaces. We also examine turbulent and laminar diffusion and premixed gas flames. A brief introduction to the Burke–Schumann problem is given and discussed, along with the calculation of the combustion rate of a thin premixed flame. The question of the combustion of solids such as thermites, plates with parallel flow and solid propellant rockets is also discussed. The appendices include a brief presentation of the Rankine–Hugoniot theory and a historical note on the research carried out in the field of combustion.

The second volume, subtitled *Case Studies and Instabilities*, is devoted to the study of the applications to one-, two- and three-dimensional spaces, gaseous and liquid physical states, supercritical fluids and single- or multi-phase systems that may be pure or mixed.

We will study the turbulence in discrete mechanics, atomization under an acceleration field, numerical simulations of tubes with sudden contraction, the evaporation of drops subjected to High Frequency disturbances, possible causes of instabilities in injection engines and flow in a Savonius wind turbine.

This series may seem a little eclectic in its study of different themes in such a vast field, and the topics are, indeed, quite diverse. However, the contents are determined by the release of articles from the OpenScience journal. We hope, of course, that other volumes will be published that will complement this diversity, and, indeed, we wish to address other themes that may be missing here, such as foam, solidification of metals or alloys, plasma in two-phase media, four-dimensional interfaces (if we consider space-time and electromagnetic fields) and so on.

Roger PRUD'HOMME
Stéphane VINCENT
Christian CHAUVEAU
Mahouton Norbert HOUNKONNOU
November 2021

1

Modeling Interfaces with Fluid Phase

It seemed necessary to reflect on interface models with at least one fluid phase. The concept and the thermodynamic and mechanical modeling of interfaces are partially revisited. The balance laws and constitutive relations are recalled, and various examples are given to support the remarks. Some questions and perspectives are also proposed.

List of symbols

Symbol	Meaning	Symbol	Meaning
A, B, C	1st, 2nd and 3rd order tensors, respectively	$\mathbf{X}$	Position vector (x, y, z) in Cartesian coordinates
C_i	Curve of coordinates	x_1, x_2, x_3	Orthogonal curvilinear coordinates
E	Internal energy	$\mathbf{X}, \mathbf{X}_{//}, \mathbf{X}_{\perp}$	Vector, its tangential, normal part
$f(\mathbf{x}, t)$	Position and time function	$\sim$	Reduced value
$\boldsymbol{f}$	Force vector acting on the unit of volume	a	Index denoting an extensive quantity per unit area, eg: ρ_a area density
F, f	Quantity, quantity F per unit mass	j	Index of a chemical species
$F(\mathbf{x}, t) = 0$	Intrinsic definition of an interface	p	Particle index

Chapter written by Roger PRUD'HOMME.

Symbol	Meaning	Symbol	Meaning
g	Thermodynamic potential per unit mass	s	Index for a quantity of interface, eg: f_S $\rho_a f_S = \int_{N_-}^{N_+} \rho f \, dN$
h	Enthalpy per unit mass	β	Reduced reaction energy $\beta >> 1$
1	Unit tensor	γ	Isentropic coefficient ratio; $\gamma = C_P / C_v$ $\gamma = (\rho^- - \rho^+)/\rho^-$
$\mathbf{J}_F, \mathcal{J}_F, \mathbf{J}_{VF}$	Flux of F for **V=0**, **V**=**v**, **V=V**	δ	Thickness of an interfacial layer
ℓ_f, ℓ_δ	Characteristic length	ε	Dimensionless number $\varepsilon << 1$
L	Phenomenological coefficient	ε_a	Surface shear viscosity
$\mathcal{L}$	Markstein length	K_a	Area viscosity
M	Mass, Mach number	μ	Shear viscosity coefficient
$\dot{m}$	Unit mass flow rate	ν	Kinematic viscosity $\nu = \mu / \rho$
n, N	Normal coordinate on a small, large scale	ξ	Coordinate matrix x_1, x_2, x_3
N	Unit normal	**Π**	Viscous pressure tensor
p,**P**	Pressure, pressure tensor (= -**Σ**)	ρ	Density
p^*	Effective pressure	σ	Surface tension
$p_{(i)}^*$	Virtual power density	**Σ**	Stress tensor
q	Heat flux vector	**σ = Σ**	Surface stress tensor
R_1, R_2	Principal curvature radii	$\boldsymbol{\sigma}_v$	Viscous stress tensor
s	Mass entropy	τ	Reduced reaction energy
S_L	Laminar flame speed	$\vartheta = 1/\rho$	Specific volume
S, S_i	Surface, coordinate surface	∇	Gradient operator
t	Time	$\nabla_{//}, \nabla_S$	Tangential gradient
T	Absolute temperature	$\nabla_\perp$	Normal gradient
V	Fluid velocity vector		

Symbol	Meaning	Symbol	Meaning
$\mathbf{V}$	Composite velocity in the interfacial layer		
$\mathbf{V}^*$	Virtual velocity		
$\mathbf{w}(\mathbf{x}, t)$	Velocity of points on a surface		
$\dot{W}_F$	F production rate		
x, y, z	Cartesian coordinates		

1.1. The concept of an interface

The general concept of an interface assumes contact or separation between material objects. While the concept of an interface in physics may refer to situations of a very different nature, there are also commonalities between these. Among these commonalities, we will study surface quantities and balance laws in particular, i.e. equations with partial derivatives that connect these quantities, in interaction with those in the media in contact.

Under certain conditions, using simplifying hypotheses, it is possible to establish balance laws for interfaces using physico-chemical quantities – constituent mass, total mass, momentum, energy, entropy – in a unique form, as is done for continuous 3D media (whether fluid or solid). We will see that this requires a small-scale internal exploration of the interface throughout its thickness, and an integration of the results obtained in the normal direction.

1.1.1. *Interface in physics and geometric surfaces*

The term "interface" refers to a separation surface. However, while the term "surface" has a precise meaning in mathematics, being a 2D manifold endowed with geometric properties, the definition of discontinuity surfaces must be refined by mechanics, more generally, in physics or chemistry.

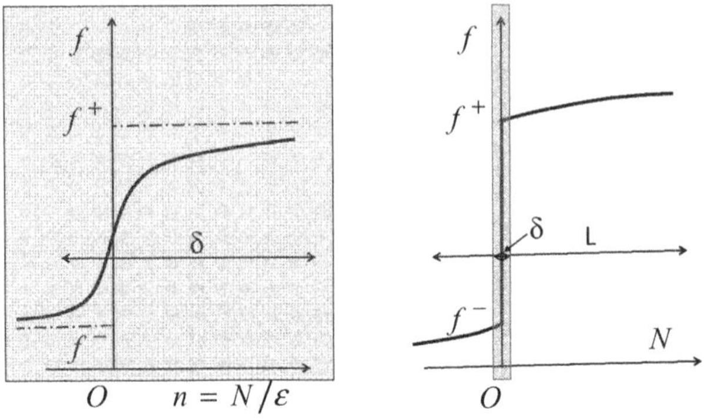

a) Interfacial layer (small scale δ) b) Interface (macroscopic scale L>> δ)

Figure 1.1. *Interfacial layer and interface. a) Gray rectangular zone obtained by dilating the X-axis of the gray part of the figure, b) the sudden growth is replaced by continuous growth between* f^- *and* f^+

Thus, as soon as this interface presents internal physical properties such as surface tension, or when it modifies the exchanges between the media that it separates, or again, when it is the site of production of different natures, it is no longer a simple separation surface.

On the contrary, there is the question of the scale of observation. The interface appears as a discontinuity surface on a macroscopic scale, but may become a continuous medium on a smaller scale. Moreover, if we observe it on an even smaller scale, the continuity gives way to atomic and molecular discontinuities and their constituents, the elementary particles (Rocard 1933; Roussel 2016).

For a mechanical physicist, the macroscopic description is required to model a problem. However, the mechanical physicist must sometimes go down to a very small scale within the objects to understand their behavior.

This is true for both interfaces and bulks. The mechanical physicist and thermodynamician willingly explore the molecular level to understand the behavior of gases, liquids or solids. The application of the laws of mechanics to objects at a molecular level makes it possible to establish macroscopic constitutive laws of mechanics for continuous media through the statistical theories that use approximations. These scientists commonly use Boltzmann's equation, Fermi's theory and the BBGKY hierarchy. The experimental measurements provide the data that allows them to use the modeling techniques developed from these laws.

It is the same for interfaces. These interfaces can be considered as surfaces with physical properties on a macroscopic scale, but if we examine them on a smaller scale, using powerful microscopes, we will see many differences. Whether or not we go as far as the molecular level, there is indeed a thickness to what appears to be a simple geometric surface. Thus, it is useful to talk about interfacial thickness and an *interfacial zone* – or an *interfacial layer* – when we study the interior of an interface in detail (Gatignol and Prud'homme 2001). This exploration will be carried out on a single scale, that of thickness, i.e. in the direction normal to the surface. Thus, the scales in the other directions that are tangential to the surface are left unchanged.

Figure 1.1 shows the classic case of an interfacial zone of very small thickness δ, compared to the characteristic length L of the other observed physical phenomena.

The concept of a continuous medium then comes into play. It is not always necessary to go as far down as the molecular level (which, by definition, is not very continuous) to model what happens in an interfacial zone. For example, capillary tension can be explained by considering that the interfacial zone is a continuous medium with properties that can be described from the second gradient theory. This does not mean that we should not go to an even smaller scale to determine the values of internal coefficients, but it is not necessary if we are only interested in the form of constitutive laws.

This is also true for volume. Indeed, the laws derived from the first gradient theory lead to the expression of the stress tensor. However, this first gradient is not sufficient to reveal interfacial behaviors, especially capillary tension. For this, we must go to the second gradient.

It should be recalled here that the virtual power method applied to continuous media postulates that the virtual power of internal efforts exerted in a control volume $(\mathcal{V})$, is the volume integral of a function $p_{(i)}{}^*$ that is a linear form of virtual speed $\mathbf{V}^*$ and its derivatives (Germain 1987). In the first gradient theory, we are limited to: $p_{(i)}{}^* = \mathbf{A} \cdot \mathbf{V}^* + \mathbf{B} : \nabla \otimes \mathbf{V}^*$, where $\mathbf{A}$ and $\mathbf{B}$ are first-order tensor coefficients. In the second gradient theory, we add a term with a third-order tensor coefficient, such that the virtual power of volume is written as: $p_{(i)}{}^* = \mathbf{A} \cdot \mathbf{V}^* + \mathbf{B} : \nabla \otimes \mathbf{V}^* + \mathbf{C} \vdots \nabla \otimes \left(\nabla \otimes \mathbf{V}^* \right)$.

1.1.2. *The concept of equilibrium in the domain of interfaces*

Phase separations can be observed at thermodynamic equilibrium at rest. For example, an interface between a liquid and a gas can present itself as being at equilibrium. We then consider its capillary tension, as well as the equilibrium of the liquid in the presence of its vapor. Similarly, thermodynamic equilibrium is a reference for solid–liquid systems, as well as for solid phases when they are multiple. The presence of stresses that are external to the system, of a mechanical or thermal nature, can lead to lesser or greater deviations from the equilibrium. The concepts of a local state and local equilibrium are used to study these systems, which are the domain of thermodynamics of irreversible processes (Defay and Prigogine 1946; Defay 1949; Prigogine and Defay 1949; Defay *et al.* 1972). However, there are systems with interfaces – which we will call *generalized interfaces* – that offer no reference to the equilibrium state. This is the case with thin flames or shock waves, for instance, which can be localized as discontinuity surfaces at a macroscopic level. However, these interfaces can also be studied when not in thermodynamic equilibrium by starting with the concept of a local equilibrium. Although these situations do not necessarily require interface laws, it is satisfying to observe the similarity in the analysis and writing between these and the case of phase separation interfaces that are out of equilibrium.

1.2. Some examples of interfaces

1.2.1. *Fluid phase change and separation interface*

This is the separation between a liquid and a gas – which can be or contain the vapor of the liquid – or between two immiscible liquids at rest.

Figure 1.2 illustrates a classic situation where the separation surface is curved because of surface tension (or capillary tension) in the presence of a cylindrical wall that constitutes a limiting condition for the interface. The meniscus is the line of contact.

Figure 1.2. *Capillary interface. Capillary surface at rest between a liquid and a gas above it, both contained in a graduated glass tube. For a color version of this figure, see www.iste.co.uk/prudhomme/fluid1.zip*

On the other hand, a moving interface with deformation, resulting from a high-velocity, annular injection of a liquid in a gas, reveals unstable motion (see (Marmottant and Villermaux 2004, Figure 7, p.79)). Apart from the capillary interface, the presence of liquid digitations (also called fibers, filaments or ligaments) can be observed.

1.2.2. *Solidification interface*

Solidification interfaces are the site of surface tensions. The phase field method is an original and productive method to study the oriented growth of eutectics (Karma and Sarkissian 1996).

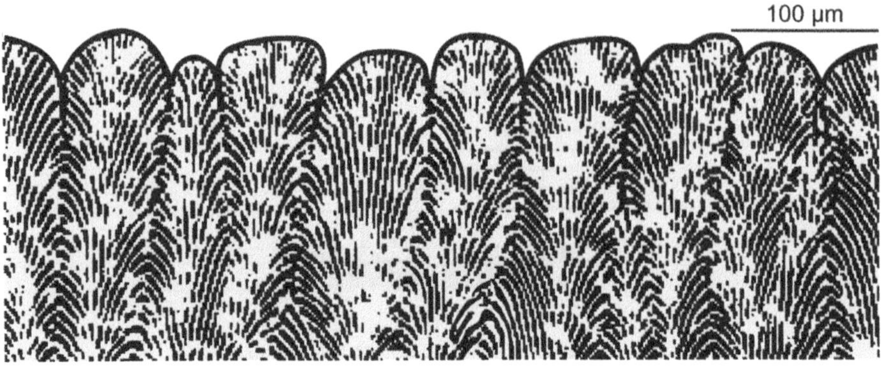

Figure 1.3. *Eutectic colonies in CBr_4-C_2Cl_6-naphthalene, $V = 31$ μm s^{-1} (Akamatsu and Faivre 2000, Figure 1.2, p. 3758)*

The phase field methods make it easier to resolve interface problems, by more or less artificially smoothening the crossing of discontinuities. In their recent form, these methods make it possible to take crystalline anisotropies into account (Akamatsu *et al.* 2015).

Phase field methods also help compute the growth of pure bodies in the presence of impurities, which can produce thick interfaces containing dendritic structures. The scales of the analysis considered depend on the problem to be resolved and lead to the use of specific phase functions.

1.2.3. *Generalized interfaces*

Zones with a high gradient allow the appearance of surfaces, even without a phase change.

Figure 1.4(a) shows a thermal drop that is of the same nature as the liquid that contains it, but at a much lower temperature and therefore at a higher density, which explains its downward motion.

Figure 1.4(b) shows a stained shock in front of a satellite during its re-entry into the atmosphere.

Figure 1.4(c) shows an observation of a premixed flame under low pressure. The surface above it is the very thin reaction zone and, the surface below it is an isotherm in the unburned mixture. At normal pressure, both of these surfaces and the zone that separates them can be combined into a single interface, which makes up the flame.

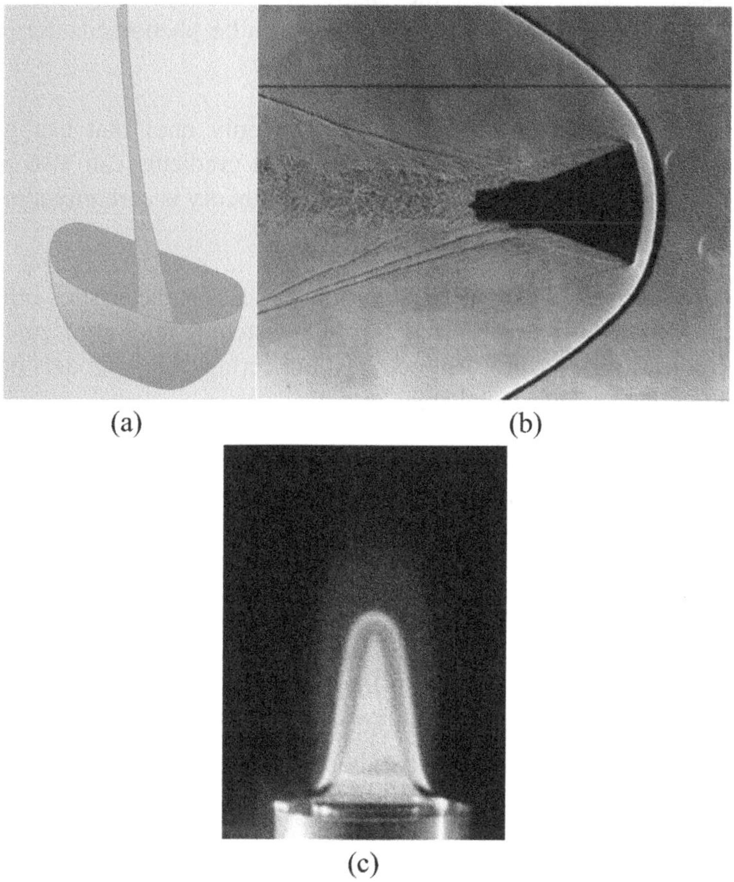

Figure 1.4. *Examples of generalized interfaces. a) Thermal drop in a liquid. b) Shockwave in front of a body re-entering the atmosphere. c) Premixed flame under low pressure, observed using a laser plane. The lower, lighter area is delimited by an isothermal surface (the temperature of the seeding oil drops, which evaporate through the heat released at the level of the reaction surface) (Prud'homme 2012). For a color version of this figure, see www.iste.co.uk/prudhomme/fluid1.zip*

Miscible liquids, brought into contact in their pure states, produce interfaces whose thickness increases over time with the mutual diffusion of the fluids. The experimental study of these interfaces reveals the existence of an effective surface tension. While it is certainly lower than the capillary tension between immiscible liquids, it is non-negligible. Such surface tension probably also exists in pure fluids subject to high thermal gradients. This would certainly explain the shape of a thermal drop. Calculations carried out on supercritical fluids show that similar phenomena are possible during the propagation of a thermal field.

Thus, high density gradients are not the only ones that can generate capillary tensions – concentration and thermal gradients can also produce effective surface tensions. The second gradient theory is certainly a means of explaining these behaviors.

With respect to shock and deflagration waves, characteristics other than surface tension are also manifested. Surface stretching rates are elements that must be considered. In turbulent combustion, certain flame models bring the volume fractions of interfacial areas into play.

Mathematical tools are required to study interfaces and interfacial zones. Modeling these systems often requires simplifying hypotheses.

1.3. Mathematical description of an interfacial layer

In reality, there are, in fact, multiple structures of interfacial and membrane layers. We have chosen to limit ourselves to layers that can be described by continuous families of surfaces that can be deformed over time (S). The interfacial layer (dilated) is bordered on either side by two specific surfaces of this family: a lower surface $\left(S^{-}\right)$ that separates it from the continuous medium below, and an upper surface $\left(S^{+}\right)$ that separates it from the continuous medium above[1]. The laws of state and, more generally, the constitutive laws of the medium of the interfacial layer and the laws of the adjacent volumic media can be similar or very different.

The present modeling is carried out by acknowledging the balance laws of the physical properties and the constitutive laws of each material medium. However, the main objective will be to establish interface laws by passing from the microscopic description of the interfacial layer, which has a certain

1 The choice of the terms "above" and "below" is arbitrary.

thickness, to the macroscopic description of the interface, which is a surface without thickness.

This operation thus involves a change in scale and an integration throughout the thickness of the interfacial zone.

However, it is possible to describe the interfacial zones in curvilinear coordinate systems, where the continuous families of surfaces (S) that are deformable over time will be coordinate surfaces. This can be considered as a real meshing of the interfacial zone – recall that this is dilated in thickness – which can be used to numerically solve the problem, but will mainly be used here for analytical purposes.

In the field of the numerical simulation of fluid mechanics governed by Navier–Stokes equations, orthogonal meshes associated with "finite differences" methods are often used. Indeed, in a large number of problems, walls are represented by curves that constitute essential information. Thus, it would be dangerous to try and account for this information through a simple succession of "staircase steps", so much so that we are naturally led to using curvilinear orthogonal meshes. These can be done by using a conformal analytical transformation. In the general case, where the shape of the walls is numerically defined, a specific program develops the orthogonal mesh that will be used to compute the flow, and defines all of the elements of the corresponding metric (Huffenus 1969). Figure 1.5 gives some examples of such meshes (Renaud-Assemat 2011). The meshes are generated for planar or revolution 2D flows as unicity problems in the case of 3D calculations.

Let us recall that these are external or internal calculations with fixed or mobile limits that are deformable, such as the surfaces of bubbles, drops or contact surfaces, often modeled by spline functions. In this regard, we must mention the work of Ryskin and Leal (1983, 1984), Duraiswami and Prosperetti (1992) and Kervella *et al.* (2012).

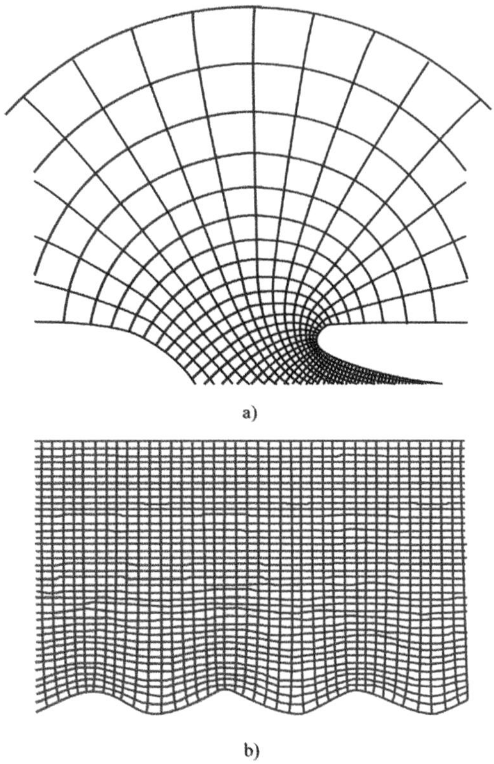

Figure 1.5. *Orthogonal meshes. a) An example of a curvilinear coordinate system that is numerically defined to calculate a confluence flow. Extracts from Huffenus (1969). b) Development of a Lagrangian method for the simulation of turbulent flows with separate phases. Extracts from Renaud-Assemat (2011)*

The methods used can nonetheless be transposed to interfacial zones by making adjustments concerning the definition of limits, which are often unknown, *a priori*, and the 3D character that was already indicated.

1.3.1. *Normal gradient and tangential gradient*

Let $f(\mathbf{x},t)$ be a function that is continuous and derivable, taking values at any point in a volume (V) and at any instant t. We can consider, in a given Cartesian location, with the coordinates $x, y, z,$ the partial derivatives of f with respect to space and time: $\partial f/\partial t$ and $\partial f/\partial x, \partial f/\partial y, \partial f/\partial z$ forming the gradient vector of f (usually denoted by $\mathbf{grad}(f)$ or ∇f, pronounced as nabla f).

The vector $\mathbf{N}$ denotes the unit normal to a surface (S) and $f(\mathbf{x},t)$, a function of space and time, taking values at any point on the surface (S) in the volume (V). The orientation of this normal is *a priori* arbitrary. It is sometimes determined by the physics of the problem.

Let us accept the existence of the partial derivatives of f at any point in (S). As in the volume, we find the partial derivatives $\partial f/\partial t$ and ∇f, but the gradient vector can be decomposed into a normal component and a tangential component.

The normal gradient of f is written as $\partial f/\partial N = \nabla_{\perp} f = (\mathbf{N}\cdot\nabla f)\mathbf{N}$ and the tangential gradient or parallel gradient can be defined as $\nabla_{//} f = (\mathbf{1} - \mathbf{N}\otimes\mathbf{N})\cdot\nabla f$, where $\mathbf{1}$ is the unit tensor. Of course, we find $\nabla f \equiv \nabla_{//} f + \nabla_{\perp} f$.

These definitions are valid for a tensor of any order of the function $f(\mathbf{x},t)$, especially if f is a scalar, a vector or a second-order tensor.

In particular, we can consider the field of unit normals $\mathbf{N}(\mathbf{x},t)$ to the surface (S), defined from the surface equations, as a function $f(\mathbf{x},t)$.

With $\mathbf{X}$ being a vector and $\mathbf{N}$ the unit normal vector to the surface, we use the following notations:

$$\mathbf{X}_{//} = (\mathbf{1} - \mathbf{N}\otimes\mathbf{N})\cdot\mathbf{X}, \quad \mathbf{X}_{\perp} = (\mathbf{N}\otimes\mathbf{N})\cdot\mathbf{X}, \quad X_{\perp} = \mathbf{N}\cdot\mathbf{X} \qquad [1.1]$$

1.3.2. *Structure and kinematics of an interfacial layer*

Let φ be a particular physical property, defined at any point in the volume of an interfacial layer. In the present model, it is accepted that the interfacial layer can be described as a set of surfaces $(S)_{\varphi}$, where φ is constant. On each surface $(S)_\varphi$ in the family, φ is constant and $\nabla_{//}\varphi$ is null. However, φ varies from one surface to another and $\nabla_\perp\varphi$ is thus non-zero[2].

Indeed, it must be accepted that the surface (*S*), in motion, is part of a continuous family of surfaces defined in a volume that contains the interfacial layer of thickness δ. On the scale L, however, the interface can be represented as a single surface.

The interfacial layer will be between the two surfaces of the families, denoted by $(S^-)_\varphi$ and $(S^+)_\varphi$.

The field of surfaces can be defined a) intrinsically, in Cartesian variables using a function *F*(*x*,*y*,*z*,*t*), or b) in orthogonal, curvilinear coordinates.

– In case a), we consider a basic surface (S_0), with the equation $F(\mathbf{x},t)=0$ and over a certain thickness on either side, a family of parallel surfaces (S_λ), defined by:

$F_\lambda(\mathbf{x},t)=F_0(\mathbf{x}+\lambda\mathbf{N},t)$; $F_\lambda(\mathbf{x},t)=0$, with $\mathbf{N}=\nabla F_0/|\nabla F_0|$ and λ being constant between $-\varepsilon$ and $+\varepsilon$.

– In case b), there is a family of surfaces with coordinates (S_3) at $x_3 = const$ (see Figure 1.6) in the system, defining the orthogonal curvilinear coordinates:

$$\begin{cases} x = x(x_1, x_2, x_3, t) \\ y = y(x_1, x_2, x_3, t) \\ z = z(x_1, x_2, x_3, t) \end{cases} \qquad [1.2]$$

2 Let us take a capillary interface, for example: on the scale L it is a capillary surface, while on a microscopic scale δ it is possible to consider the local density ρ and S_ρ, with ρ being constant.

or again, $\mathbf{x} = \mathbf{x}(x_1, x_2, x_3, t) = \mathbf{x}(\xi, t)$. We set $\mathbf{h}_i = \dfrac{\partial \vec{\mathbf{x}}}{\partial x_i}$, $|\mathbf{h}_i| = h_i$, $\mathbf{h}_i = h_i\, \mathbf{e}_i$ and we have $\mathbf{e}_i \cdot \mathbf{e}_j = \delta_{ij}$.

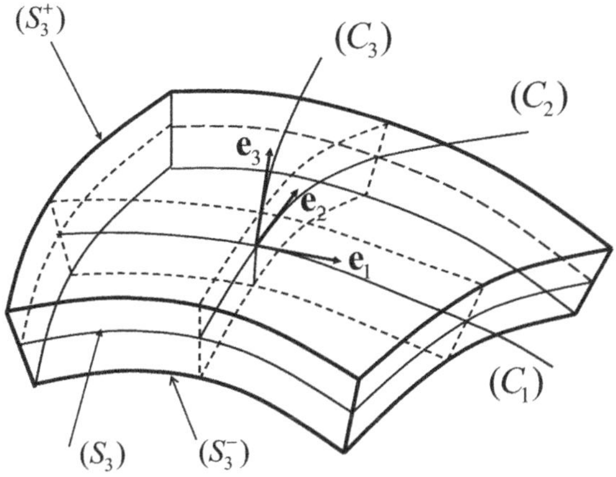

Figure 1.6. *Coordinate surfaces and interfacial layer. (C_i) are the coordinate curves*

The family thus defined depends on time. The set of surfaces that it makes up moves through deformation. In case b), the surfaces in the family are not necessarily parallel. They are only parallel (and remain this way) if the condition $\partial \mathbf{e}_3 / \partial x_3 = 0$ is verified at all instants, where $\mathbf{e}_3$ is the unit vector of the curvilinear coordinate x_3 (see Figure 1.6). In case a), we must take λ as the position function if the surfaces are not required to be parallel.

Let us now apply the derivation operators defined in section 1.3.1. With $f = \mathbf{N}$, it can be shown that $|\nabla \cdot \mathbf{N}|$ is equal to the mean curvature $(1/R_1 + 1/R_2)$ of the surface at the point and instant being considered.

Let us now consider time t. It can be assumed that any point on a surface (S) is activated by a speed $\mathbf{w}(\mathbf{x}, t)$. It is then possible to arrive at the rate of deformation of this surface, assumed to consist of points with the speed $\mathbf{w}$. The normal and tangential speeds are defined as $\mathbf{w}_\perp = w_\perp \mathbf{N}$ (with $w_\perp = \mathbf{w} \cdot \mathbf{N}$) and $\mathbf{w}_{//} = (\mathbf{1} - \mathbf{N} \otimes \mathbf{N}) \cdot \mathbf{w}$, respectively. This leads, in particular, to the *stretching rate* $\nabla_{//} \cdot \mathbf{w}$ of the surface (S) defined. This definition of the surface velocity comes from mechanics, which is based on the mechanics of a material point.

With respect to the geometric aspect, only the normal velocity $\mathbf{w}_{\perp}$ is defined, knowing $F(\mathbf{x},t)$. From the intrinsic definition, we find $w_{\perp} = -\dfrac{\partial F(x,y,z,t)/\partial t}{|\nabla F|}$.

However, if the points on the surface are defined by $\mathbf{x} = \mathbf{x}(\xi,t)$ with $\xi = (x_1, x_2, x_3)$, the velocity of these points is thus at t: $\mathbf{w} = \partial \mathbf{x}(\xi,t)/\partial t$. As was done earlier, $\mathbf{w}_{\perp}$ can be considered; this is the velocity at which the geometric surface advances, which must be distinguished from the tangential sliding movement of the interface on itself, characterized by $\mathbf{w}_{//}$.

Studying the interior of the interfacial layer implies knowing its structure. This can be more or less curved and of a non-homogeneous thickness. It will be chosen based on the knowledge of the interfacial physico-chemistry, which is not always precisely known.

1.3.3. *Bulk and surface quantities*

We must review the extensive quantities – and their mass and volume values – used in the volume (including that of the interfacial layer) and define the corresponding surface quantities.

The main extensive bulk values F are the total mass, M, the masses, M_j of the constituent elements (in the case of a mixture), the volume, $\mathcal{V}$, the momentum, $M\mathbf{v}$, the internal energy, E, the kinetic energy, K, and the entropy, S. We use the density of the mixture, ρ (symbolically $M/\mathcal{V}$), and the densities of the species, $\rho_j = M_j/\mathcal{V}$. The quantities per unit mass $f = F/M$ are $\mathbf{v}, e, k, s$ (respectively the barycentric velocity vector, the internal energy, the kinetic energy and entropy per unit mass). We also define the flux vector $\mathbf{J}_{VF}$ and the production rate $\dot{W}_F$.

The area density ρ_a is defined by $\rho_a = \int_{N_-}^{N_+} \rho dN$. This makes it possible to define the surface value f_S corresponding to the quantity F: $\rho_a f_S = \int_{N_-}^{N_+} \rho f \, dN$, as well as the flux and the surface production: $\mathbf{J}_{VFa} = \int_{N_-}^{N_+} \mathbf{J}_{VF} \, dN$, $\dot{W}_{Fa} = \int_{N_-}^{N_+} \dot{W}_F \, dN$. Note that we must distinguish between the indices a and S. Indeed, the first index addresses the quantities per unit area, while the second simply signifies that the quantity considered is a local weighted average at the interface, i.e. crossing through the interfacial zone.

1.3.4. *Interface balances*

Consider an extensive quantity F, such as the mass M_j of the species j in the mixture, the total mass $M = \sum_j M_j$, the momentum – partial, $M_j \mathbf{v}_j$, or total, $M \mathbf{v} = \sum_j M_j \mathbf{v}_j$ –, the internal energy E or the entropy S.

We start from the volume balance equation of the quantity F:

$$\frac{d(\rho f)}{dt} + \rho f \nabla \cdot \mathbf{v} + \nabla \cdot \mathcal{J}_F = \dot{W}_F \qquad [1.3]$$

where $f = F/M$ is the local quantity per unit mass, $\mathcal{J}_F$ is the barycentric unit flux and $\dot{W}_F$ is the production rate of F per unit volume.

By introducing the unit flux across a fixed surface $\mathbf{J}_F = \mathcal{J}_F + \rho f \, \mathbf{v}$, another form of the local balance equation is obtained:

$$\frac{\partial(\rho f)}{\partial t} + \nabla \cdot \mathbf{J}_F = \dot{W}_F \qquad [1.4]$$

If a unit flux is now introduced across a surface, relative to a motion with velocity $\mathbf{V}$:

$\mathbf{J}_{VF} = \mathcal{J}_F + \rho f (\mathbf{v} - \mathbf{V})$, the following form is obtained:

$$\frac{d_{\mathrm{V}}(\rho f)}{dt}+\rho f \nabla\cdot\mathbf{V}+\nabla\cdot\mathbf{J}_{\mathrm{V}F}=\dot{W}_F \qquad [1.5]$$

Interface balance laws can be established by reshaping the 3D balance equation (interfacial layer, Figure 1.6), which can be written as:

$$\frac{d_{\mathrm{V}}(\rho f)}{dt}+\rho f \nabla_{//}\cdot\mathbf{V}+\nabla_{//}\cdot\mathbf{J}_{\mathrm{V}F}+\frac{\partial J_{\mathrm{V}F\perp}}{\partial N}=\dot{W}_F \qquad [1.6]$$

where $\mathbf{J}_{\mathrm{V}F}=\mathcal{J}_F+\rho f(\mathbf{v}-\mathbf{V})$, $\mathbf{J}_{\mathrm{V}F}=\mathbf{J}_{\mathrm{V}F//}+\mathbf{J}_{\mathrm{V}F\perp}$, and also $\mathbf{J}_{\mathrm{V}F//}=(\mathbf{1}-\mathbf{N}\otimes\mathbf{N})\cdot\mathbf{J}_{\mathrm{V}F}$ and $\mathbf{J}_{\mathrm{V}F\perp}=J_{\mathrm{V}F\perp}\mathbf{N}$

which yields $\nabla_{//}\cdot\mathbf{J}_{\mathrm{V}F}=\nabla_{//}\cdot\mathbf{J}_{\mathrm{V}F//}+J_{\mathrm{V}F\perp}\nabla\cdot\mathbf{N}, \nabla_{\perp}J_{\mathrm{V}F\perp}=\partial J_{\mathrm{V}F\perp}/\partial N$.

The next step consists of integrating the two sides of the previous bulk balance equation over the thickness of the interfacial layer. Note that it is assumed that some quantities are preserved all along the normal curve to the surfaces (*S*), starting from the point P under consideration (Prud'homme 2010). It is also assumed that these surfaces are parallel to each other. These hypotheses are verified – especially concerning the velocity vector $\mathbf{V}$ – for both types of interfacial zones shown in Figure 1.7. The following interface balance equation is obtained:

$$\frac{d_S(\rho_a f_S)}{dt}+\rho_a f_S\nabla_S\cdot\mathbf{V}_S+\nabla_S\cdot\mathbf{J}_{VFa}+[J_{VF\perp}]_-^+=\dot{W}_{Fa} \qquad [1.7]$$

with the definitions given earlier (section 1.3.3) for the indexed interfacial quantities *a* and *S*.

The similarity with the bulk balance equation [1.5] can be seen, with the exception of one term of jump. This term is equal to:

$[J_{\mathrm{V}F\perp}]_-^+=[\mathbf{J}_{\mathrm{V}F}{}^+-\mathbf{J}_{\mathrm{V}F}{}^-]\cdot\mathbf{N}$, and characterizes the algebraic increase in the normal flow of the quantity *F* during the crossing of the interface, following the movement of the velocity $\mathbf{V}$.

The general balance equation of the interface is applied to the various quantities considered, and from this we can derive balance equation tables

(Slattery 1990; Gatignol and Prud'homme 2001; Prud'homme 2012), which are reproduced below for a classic mixture.

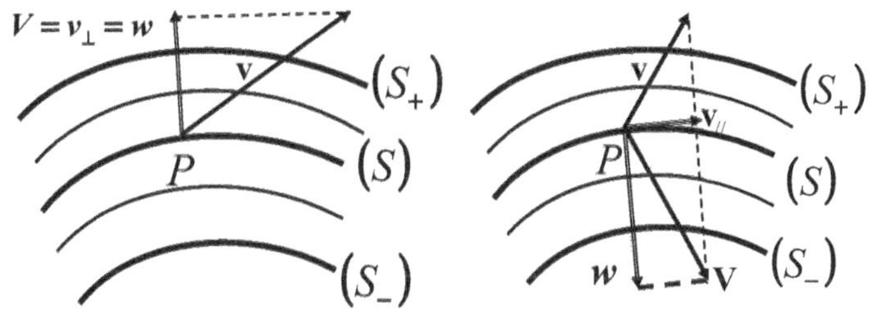

a) Interfacial zone of type "Shear layer"

b) Interfacial zone of type "Shock wave"

Figure 1.7. *Definition of the velocity vector* **V** *for two common types of interfacial layer (planar representation)*

Continuity equation of the species	$\dfrac{d_S(\rho_a Y_{jS})}{dt} + \rho_a Y_{jS} \nabla_S \cdot \mathbf{v}_S + \nabla_S \cdot \mathcal{J}_{Dja} + \left[\mathcal{J}_{Dj\perp} + \rho_j u_\perp\right]_-^+ = \dot{W}_{ja}$
Continuity equation	$\dfrac{d_S \rho_a}{dt} + \rho_a \nabla_S \cdot \mathbf{v}_S + \left[\rho u_\perp\right]_-^+ = 0$
Equation for the momentum	$\dfrac{d_S(\rho_a \mathbf{v}_S)}{dt} + \rho_a \mathbf{v}_S \nabla_S \cdot \mathbf{v}_S - \nabla_S \cdot \boldsymbol{\sigma}$ $+\left[\mathbf{P} \cdot \mathbf{N} + \rho(\mathbf{v} - \mathbf{v}_S) u_\perp\right]_-^+ = \sum_j \rho_a Y_{jS} \mathbf{f}_{jS}$
Energy equation	$\dfrac{d_S(\rho_a e_S)}{dt} + \rho_a e_S \nabla_S \cdot \mathbf{v}_S + \nabla_S \cdot \mathbf{q}_a$ $+\left[q_\perp + \rho e u_\perp\right]_-^+ = \sum_j \mathcal{J}_{Dja} \cdot \mathbf{f}_{jS}$ $+\boldsymbol{\sigma} : \nabla_S \otimes \mathbf{v}_S - \left[\dfrac{1}{2}\rho(\mathbf{v} - \mathbf{v}_S)^2 u_\perp + (\mathbf{v} - \mathbf{v}_S) \cdot \mathbf{P} \cdot \mathbf{N}\right]_-^+$
Entropy equation	$\dfrac{d_S(\rho_a s_S)}{dt} + \rho_a s_S \nabla_S \cdot \mathbf{v}_S + \nabla_S \cdot \mathcal{J}_{sa} + \left[\mathcal{J}_{S\perp} + \rho s u_\perp\right]_-^+ = \dot{W}_{Sa}$

Table 1.1. *Balance equations for interfaces whose surface variables obey classic thermodynamic relations, with* $\mathbf{V}_s = \mathbf{v}_s, u_\perp = v_\perp - v_{S\perp}$ *and* J_{Dj}, *j being the diffusion flux of the species j*

1.3.5. *Constitutive laws*

The various quantities that come into play in the equations given in Table 1.1 are not independent of one another.

First of all, there are all of the constitutive laws of the media in contact with the interface: first, the equations of state and then the complementary constitutive laws. The thermodynamic quantities obey the equations of state of fluid mixtures and are assumed to verify the Gibbs and Gibbs–Duhem relations. The internal energy e per unit mass is a function of the entropy s per unit mass and the mass density ρ, and we have $e = e(s, \rho)$, which verifies the Gibbs relation $de = T\,ds - p\,d(1/\rho)$.

The other media in contact – which can be solids, as in the case of the solidification of alloys – verify thermodynamic relationships that are specific to them. Complementary constitutive laws govern dissipative processes: viscosity, thermal conduction, diffusion of species and chemical reactions. It is assumed that the constitutive relations of the continuous media in contact with the interface are known.

The indexed interface quantities a or S pose other problems, and their relations are often required to be determined. Several situations arise depending on whether or not the local properties, thermodynamic and mechanical, of the interfacial layer are known:

– If the local internal properties of the interfacial layer are known, it is possible to calculate what happens through integration along the normal, as formally done in sections 1.1.1 and 1.3.3 to obtain the equations in Table 1.1. In this way, the generalized interfaces such as discontinuity waves, boundary layers or mixing layers can be studied. Certain tricks often make it possible to avoid complex calculations by analyzing the evolutions on the scale L, without having to use an analysis on the scale δ. Or again, it is possible to stop with asymptotic connections between the scales.

– If the local properties on a small scale $\delta << L$ are not known, we are led to make hypotheses that must be verified *a posteriori* through calculations or experimentation.

It must be noted that the capillary interfacial layers are not conventional fluids, but fluids endowed with internal capillarity. It can be demonstrated that a conventional fluid interfacial layer does not lead to an interface with capillary tension. The capillary tension between immiscible fluids only arises from this condition. Thus, the internal energy must be written as $e = e\left(s, \rho, |\nabla\rho|^2\right)$. The new Gibbs relation becomes $de = T\,ds - p\,d(1/\rho) + \lambda(\nabla\rho)\cdot d(\nabla\rho)$, where λ is the internal capillarity coefficient. Seppecher determined the relationship between the internal capillarity coefficient and surface tension (Seppecher 1987):

$$\sigma = -\int_{N^-}^{N^+} p^*(\xi)\,d\xi \cong \int_{N^-}^{N^+} \lambda|\nabla\rho|^2\,d\xi \qquad [1.8]$$

The method used is that of the second gradient (Gatignol and Prud'homme 2001).

In addition to a scalar thermodynamic capillary tension σ, a capillary interface in motion can also present a mechanical surface tension tensor for the components tangential to the surface: $\boldsymbol{\sigma} = \boldsymbol{\sigma}_{//}$.

Slattery (1990) proposed the following expression, given in Table 1.2, for the viscous part of the surface tension of a moving interface with a capillary tension tensor:

$$\boldsymbol{\sigma}_v = \boldsymbol{\sigma} - \sigma\mathbf{1}_{//} = \left[(\kappa_a - \varepsilon_a)\nabla_S\cdot\mathbf{v}_S\right]\mathbf{1}_{//} + 2\varepsilon_a\mathbf{D}_S \qquad [1.9]$$

with $\mathbf{D}_S = \frac{1}{2}(\nabla_S\cdot\mathbf{v}_S)\mathbf{1}_{//} + \boldsymbol{\Delta}_S$, where $\boldsymbol{\Delta}_S$ is a 2D tensor with zero trace and the coefficients of the shear viscosity coefficient and the area, respectively, ε_a and κ_a, are real, non-negative numbers according to the second principle of thermodynamics. These surface viscosities are notably a function of the stretching rate $\nabla_S\cdot\mathbf{v}_S$.

In other cases, we will try to directly establish constitutive laws by analyzing the expression of the production of interfacial entropy. Let us recall the expression for the rate of production of interfacial entropy obtained for conventional reactive fluid mixtures (Prud'homme 2012), from the equations given in Table 1.1 and from the Gibbs relation:

$$\dot{W}_{Sa} = -\sum_j \left(g_{jS}/T_S\right)\dot{W}_{ja} + \left(1/T_S\right)\boldsymbol{\sigma}_v : \nabla_S \otimes \mathbf{v}_S + \mathbf{q}_a \cdot \nabla\left(1/T_S\right) - \sum_j \mathbf{J}_{ja}\left[\nabla_S\left(g_{jS}/T_S\right) - \mathbf{f}_j/T_S\right]$$
$$+ \left[\left(q_\perp + \rho h u_\perp\right)\left(1/T - 1/T_S\right)\right] - \sum_j \left(J_{Dj\perp} + \rho\left(Y_j - Y_{jS}\right)u_\perp\right)\left(g_j/T - g_{jS}/T_S\right) \qquad [1.10]$$
$$- \left(1/T_S\right)\left(\mathbf{v} - \mathbf{v}_S\right)\cdot \boldsymbol{\Pi}\cdot \mathbf{N} + \left(1/2T_S\right)\rho\left(\mathbf{v} - \mathbf{v}_S\right)^2 u_\perp]_-^+ \geq 0$$

This expression for the rate of production of entropy must be shaped to reveal a series of forces generalized by flows. We then identify the identical orders of tensors: 0, 1 and 2. Furthermore, by following the principles of the linear thermodynamics of conventional irreversible processes, it is possible to set the linear relations of coupling between generalized forces and generalized fluxes, assumed to be sufficiently small. Then, matrices of phenomenological coefficients with relations of symmetry appear. This is the method that Bertrand and Prud'homme attempted to use, with limited success (Bertrand and Prud'homme 1977).

However, there are also cases with nonlinear relationships between the generalized fluxes and the generalized forces that cause them. In general, this is the case with the rates of chemical production. This can also be observed for shockwaves.

1.4. Some additional information and examples of application

The following considerations aim to demonstrate the importance of the concept of an interface by widening its field of application, while relativizing its relevance. We will specifically give examples of generalized interfaces.

1.4.1. *Effective surface tension between two miscible liquids*

The second gradient theory is valid for superimposed fluid layers that do not diffuse into one another.

However, it can be experimentally demonstrated that surface tension can exist, even in the presence of mutual diffusion (Petitjeans *et al.* 1999; Pojman *et al.* 2006).

Defay and Sanfeld (1973), for example, proposed an expression for surface tension that is a function of the concentration gradients of the fluids a and b: $\sigma = e^3 \int_{N^-}^{N^+} \sum_a \sum_b k_{ab} c'_a \, c'_b \, dz$, where z is the direction perpendicular to the interface with the characteristic thickness e. The constant k_{ab}, which must be determined, is only dependent on the nature of the fluids.

We propose examining the applicability of the second gradient method to the case of miscible fluids. In this case, we propose replacing the variable $\nabla\rho$ by the gradient for the mass fraction ∇Y in the interfacial layer equations and applying the second gradient method (Gatignol and Prud'homme 2001), as in section 1.3.5.

1.4.2. *Terms that come into play in interface balance laws*

We examine different types of interfaces. Some of these only exist out of equilibrium, in the presence of strong dissipation: for example, viscous and heat-conductive dissipations for shockwaves, or heat dissipations mass diffusion and chemical kinetics that are out of equilibrium for deflagration waves.

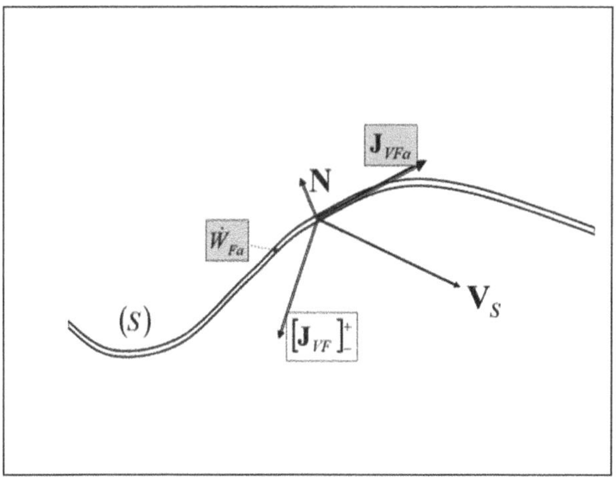

Figure 1.8. *Different terms that come into play in the general equation of interfaces [1.7], represented here for a scalar quantity F: $\mathbf{V}_S$ is the composite local speed of the interface, $\mathbf{J}_{VFa}$ is the tangential flow of the interface, $\dot{W}_{Va}$ is the rate of surface production and $[J_{VF\perp}]_-^+ = [\mathbf{J}_{VF}]_-^+ \cdot \mathbf{N}$ is the volume flux jump of F across the interface. For a color version of this figure, see www.iste.co.uk/prudhomme/fluid1.zip*

Each case verifies the general interface balance equation [1.7]:

$$\frac{d_S(\rho_a f_S)}{dt} + \rho_a f_S \nabla_S \cdot \mathbf{V}_S + \nabla_S \cdot \mathbf{J}_{VFa} + [J_{VF\perp}]_-^+ = \dot{W}_{Fa}$$

This equation was established using the approach discussed in section 1.3.4. Certain interfacial layers are made up of capillary fluids. However, sometimes the bulk equations are those of simple fluids or those in a conventional mixture of fluids, with complementary equations of state and constitutive relations, whose linearized forms are partly derived from the thermodynamics of irreversible processes (TIP) through the examination of the expression for the rate of production of entropy. The phenomenological coefficients of these irreversibilities come into play during the integration through the interface, and it is hoped that these calculations will lead to new phenomenological coefficients for the interface that characterize viscosities, thermal conduction and mass diffusion processes. Chemical production rates, usually nonlinear, will necessitate special treatment, as is done for thin premixed flames, for example, for which multiscale asymptotic expansions are required.

Figure 1.8 highlights the kinds of terms that come into play in interface balance equations.

1.4.3. *Normal shockwave*

Normal shockwaves are among the simplest examples of generalized interfaces. This problem has already been widely studied without using the concepts explained above. The formulae for stationary normal waves in a perfect gas are well known. We do not seek to add anything new to this topic but will use it as an example to find out whether our theory is applicable, in particular to determine which constitutive laws are valid for this type of discontinuity surface. For example, what can the Hugoniot relations teach us? What relations exist between the generalized forces and fluxes?

The balance equations for a normal shockwave are as follows:

$$\begin{gathered}[\rho u]_-^+ = [\dot{m}]_-^+ = 0(\text{mass}), \dot{m}[u]_-^+ + [p]_-^+ = 0 \\ (momentum), \dot{m}\left[h + u^2/2\right]_-^+ = 0(\text{energy})\end{gathered} \quad [1.11]$$

and the rate of production of entropy is reduced to $\dot{W}_{Sa} = \dot{m}\,[s]_-^+ = \dot{m}\,c_v \ln \tilde{p}/\tilde{\rho}^{\gamma}$, using $\tilde{p}$ and $\tilde{\rho}$ to denote pressure and density as the ratio of their upstream values.

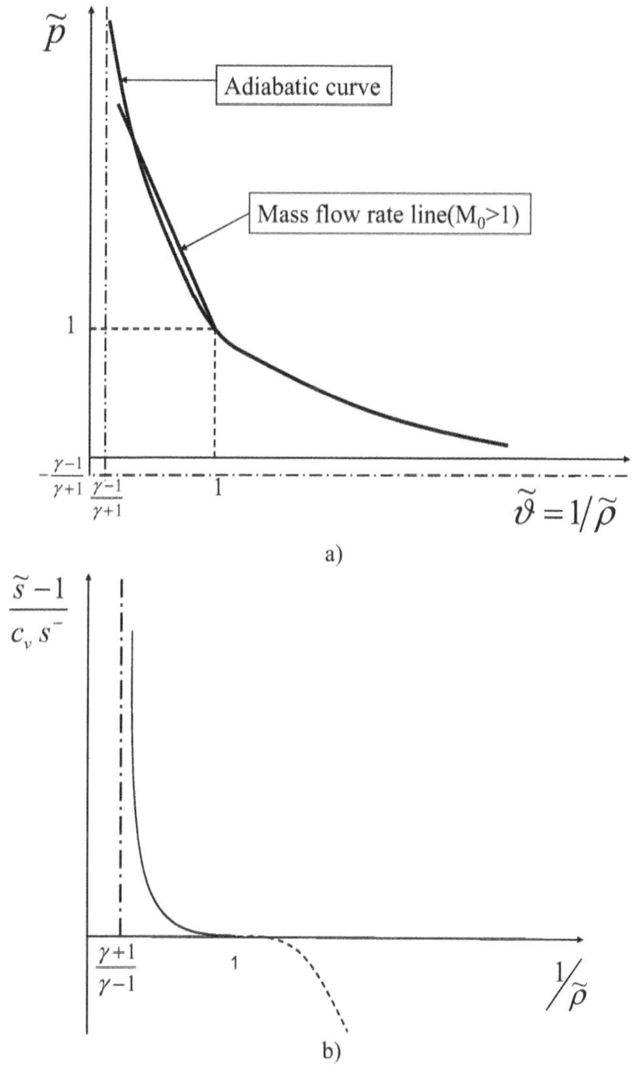

Figure 1.9. *Hugoniot adiabatic curve in the planes a) $\tilde{\vartheta},\tilde{p}$ and b) $\tilde{\vartheta},\tilde{s}$. The Mach number for the upstream flow, with respect to the wave, is $M^- = M_0 = u_0 / c_0$. A shock only exists for $M^- > 1$. The equation for the mass flow rate line is $\left(\tilde{p}-1\right)/\left(\tilde{\vartheta}-1\right) = -\gamma M_0^2$*

Figure 1.9 shows that the entropy variation curve has a point of inflection that is a horizontal tangent at the point $\tilde{\vartheta} = 1$, and that the calculation shows[3] that $\dot{W}_{Sa} = -\dot{m}\, c_v \left(\gamma^2 - 1\right)\left(\tilde{\vartheta} - 1\right)^3 / 12$. If we examine the relationship between the pressure jump (the cause of discontinuity) and the specific volume jump (the resulting generalized flow), we will not find a linear relationship as is the case in TIP, but a second-degree expression: $1 - \tilde{\vartheta} \propto (\tilde{p} - 1)^2$. The result summarized in Table 1.2 is obtained for weak shocks.

What happens if we study the interfacial zone that makes up the interior of the shock?

In this zone, it is possible to write, as proposed by Smolderen (1971), the equations for a viscous and heat-conductive fluid with linear constitutive laws. However, integration along the normal to the interface leads to a nonlinear constitutive law.

This behavior is also observed for shocks in two-phase media, as the shock involves the gaseous phase. However, in a gas–particle mixture, linear phenomenological relations arise (see Table 1.2), expressing thermal and viscous exchanges between the gas and the particles (Jamet *et al.* 1997; Jamet 1998).

1.4.4. *Combustion waves*

The Rankine–Hugoniot equations (Jaumotte 1971) characterize a flat discontinuity wave in the presence of a chemical combustion reaction. The basic equations are similar to the equations for the shockwaves:

$[\rho u]_-^+ = [\dot{m}]_-^+ = 0$, $\dot{m}[u]_-^+ + [p]_-^+ = 0$, $\dot{m}\left[h + u^2/2\right]_-^+ = 0$; however, the equations of the state of the mixture must be added to these equations. It is assumed that there is constant and uniform specific heat, and an energy Δh brought into play through the chemical reaction, such that $[h]_-^+ = c_p(T_1 - T_0) - \Delta h$.

3 A normal shock only exists for Mach M_0 greater than unity. At the point $\vartheta = 1$ in Figure 1.9, there is no shock as of yet. For a Mach number slightly greater than 1, there is a weak shock.

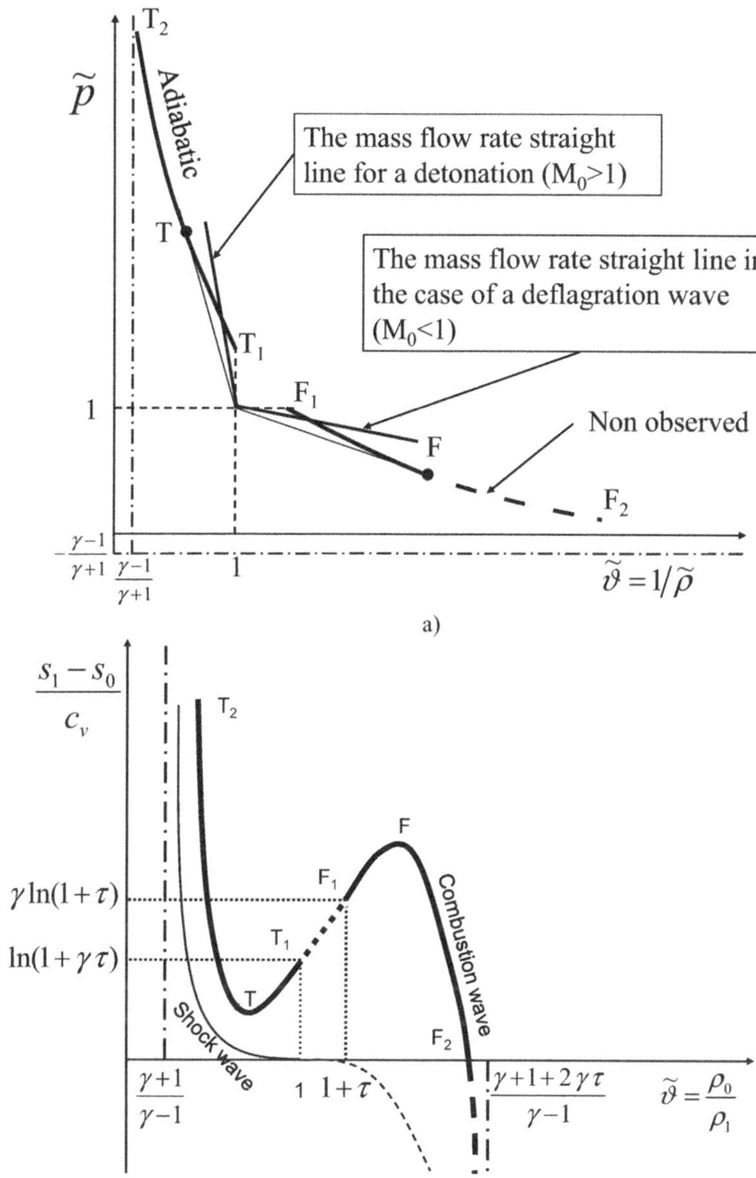

Figure 1.10. *a) The line of mass flow rate (in both cases) and the Hugoniot adiabatic curve. b) The rate of production of entropy, across the combustion wave and along the Hugoniot adiabatic curve. Arc* F_1FF_2 *is in the zone of positive variation and allows an expansion discontinuity wave. The shock represented by thin lines does not present such a wave*

When all of the calculations are completed and by setting $\widetilde{\vartheta}=\rho_0/\rho_1$, $\widetilde{p}=p_1/p_0$, $\tau=\Delta h/c_pT_0$, there are two dimensionless equations:

$$\frac{\widetilde{p}-1}{\widetilde{\vartheta}-1}=-\gamma M_0^{\ 2},\ \widetilde{p}=\frac{(1-\gamma)\widetilde{\vartheta}+1+\gamma+2\gamma\tau}{(1+\gamma)\widetilde{\vartheta}+1-\gamma} \qquad [1.12]$$

which describe, respectively, the line of mass flow rate and the Hugoniot adiabatic curve.

Two regimes appear[4]:

– The deflagration wave or premixed flame, which is subsonic and for which the pressure drop is relatively low; the thin laminar flame, planar and adiabatic, which has a combustion rate s_L^0.

– The detonation wave, which is supersonic and stable at the Chapman–Jouguet point T, has a high amplitude.

Nature of the interface	**Balance and constitutive laws**	**Method**
Normal shock wave	$[\rho u]_-^+=[\dot m]_-^+=0,\ \dot m[u]_-^++[p]_-^+=0,\ \dot m[h+u^2/2]_-^+=0$ $\dot W_{Sa}=\dot m[s]_-^+=\dot m c_v\ln\widetilde{p}/\widetilde{\rho}^{\gamma}\propto(M_0^{\ 2}-1)^3$	Conventional theory of stationary shock waves
Two-phase stationary shock	$\dot W_{Sa}\cong Q_a\left(\frac{1}{T_{pS}}-\frac{1}{T_S}\right)+\frac{F_a}{T_S}(u_S-u_{pS}),u_S=\frac{u_1+u_2}{2},u_{pS}=\frac{u_0+u_2}{2}$ $Q_a=\mathcal{L}_{TS}(T_S-T_{pS}),F_a=\mathcal{L}_{vS}(u_S-u_{pS})$	Linear TIP applied to the relaxation zone that follows the gaseous shockwave
Thin premixed flame	$s_L/s_L^0\cong 1-\mathcal{L}(\kappa+S/s_L^0)$	Asymptotic analysis with two small parameters

4 Contrary to the case of the shockwave, this time a combustion wave exists for the Mach M_0 smaller than 1, due to the presence of the exothermic reaction.

Nature of the interface	Balance and constitutive laws	Method
Boundary, incompressible stationary boundary layer above a flat plate	$\nabla_S \cdot \mathbf{J}_{ua} + [J_{u\perp}]_-^+ = 0$ $J_{ua} = -K_u \rho U_\infty x\, \psi\left(\dfrac{\mu}{\rho U_\infty x}\right)[u]_-^+$	Dimensional analysis
Incompressible non-stationary boundary layer above a flat plate	$\dfrac{\partial \rho_a \Delta u_S}{\partial t} - [\tau]_-^+ = 0$ $\rho_a \Delta u_S = L_u \rho\, d\, u [u]_-^+$	Dimensional analysis
Diffusive boundary layer above a flat plate	$\dfrac{\partial J_{Ya}}{\partial x} + [J_{Y\perp}]_-^+ = 0$ $J_{Ya} = -K_Y \rho U_\infty x \mathrm{Re}_x^{\,2} Sc^b [Y]_-^+$	Dimensional analysis
Capillary surface	$[\mathbf{P} \cdot \mathbf{N}]_-^+ - \nabla_S \cdot \boldsymbol{\sigma} = \mathbf{0}$ $\boldsymbol{\sigma}_v = \boldsymbol{\sigma} - \sigma \mathbf{1}_{//} = [(\kappa_a - \varepsilon_a)\nabla_S \cdot \mathbf{v}_S]\mathbf{1}_{//} + 2\varepsilon_a \mathbf{D}_S$	Direct TIP

Table 1.2. *Examples of the constitutive laws of interfaces*

1.4.5. *Thin premixed flame*

Let us now consider the thin premixed flame shown in Figure 1.11, which is no longer planar or adiabatic. When the activation energy of the chemical reaction is very high, the reaction zone is very thin and the flame has a well-defined structure (see also Prud'homme (1988) and Prud'homme (2012, pp. 177–179), as well as Gatignol and Prud'homme (2001) and Prud'homme (2010) for details) and includes four zones:

– fresh gases;

– the preheating zone of thickness ℓ_f, dominated by convection, the diffusion of species and thermal conductivity;

– the reactive zone of thickness ℓ_δ, where diffusion, conduction and chemical reactions are predominant and can be represented by a characterized jump – by a Dirac delta – giving the dimensionless reaction speed: $W = \delta(N)\exp \beta\left(\tilde{T}^+ - 1\right)$;

– burnt gases.

The interfacial layer is made up of the "second and third zone". The reference combustion speed s_L^0 corresponds to the adiabatic, planar, laminar flame, which is the speed of the combustion wave with respect to cool gases. The 3D theory gives the speed s_L of the curved and stretched flame (Clavin and Joulin 1983) with respect to s_L^0:

$$s_L / s_L^0 \cong 1 - \mathcal{L}\left(\kappa + S / s_L^0\right) \qquad [1.13]$$

depending on the stretching rate of the interface $\nabla_{//} \cdot \mathbf{V}_S$, decomposed into two parts: the curvature of the flame front κ and the local strain rate of the fluid flow $S = -\mathbf{N} \cdot \nabla \otimes \mathbf{v} \cdot \mathbf{N}$ divided by s_L^0.

The Markstein length $\mathcal{L}$ is defined by:

$$\frac{\mathcal{L}}{\ell_F} = \frac{1}{\gamma} \ln \frac{1}{1-\gamma} + \frac{Ze(Le-1)}{2} \frac{1-\gamma}{\gamma} \int_0^{\frac{\gamma}{1-\gamma}} \frac{\ln(1+x)}{x} dx \qquad [1.14]$$

with $\gamma = \left(\rho^- - \rho^+\right) / \rho^-, Ze = E_\alpha \left(T^+ - T^-\right) / R\left(T^+\right)^2$.

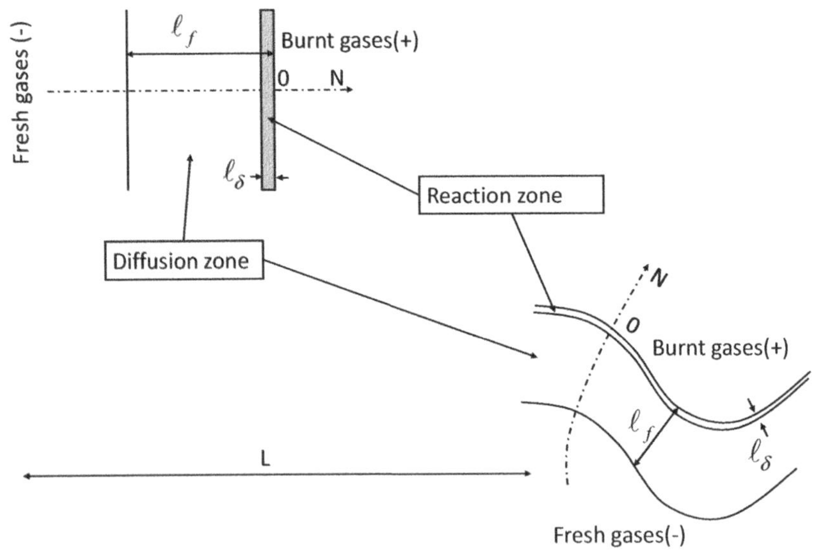

Figure 1.11. *Comparison of the structures of the planar flame and the curved flame*

The - and + sides denote, respectively, fresh and burnt gases, with *Le* being the Lewis number – assumed to be in the vicinity of unity – and *Ze* the Zeldovich number.

Interface equations can be used in the asymptotic analysis itself. Thus, it must be noted that the calculations over the asymptotic expansions relating to the 3D balance equations can also be carried out over all of the interface balance equations. The results are the same, as demonstrated in Prud'homme (1988).

1.4.6. *Boundary layers*

In problems with a discontinuity wave, the speed of the interface is that of case b) in Figure 1.7, while the boundary layers correspond to Figure 1.7(a).

In Table 1.2, the cases presented were analyzed using dimensional analysis, which make it possible to obtain constitutive laws. However, these are simple geometric configurations and 2D flows above flat plates, which may be fixed or mobile.

Other configurations may require different treatments.

1.5. Conclusion

This chapter has presented a method for establishing balance equations for interfaces in the presence of at least one fluid phase[5]. This method is based on the hypothesis that there are at least two spatial scales of analysis and that on a small scale the interface becomes a fluid layer that can have original properties. The balance equations developed have the advantage of revealing the internal exchanges and production rates at the interfaces, and the exchanges between these interfaces and the volumic media which they are in contact with.

5 Fluids that present density gradients are subject to the action of volumic forces at a distance, such as gravity. These forces and microgravity forces are taken into account in this modeling of interfaces, which is of particular interest for studies carried out in the GDR MFA framework (Fundamental and Applied Microgravity on systems that present a fluid phase).

The equations that are established contain new quantities that are generally not independent of one another, but obey constitutive laws. It has been observed that these laws can be established in a variety of ways depending on the configurations of the interfaces being considered.

Certain fundamental questions remain. We discussed diffusion interfaces, for which the appearance of an effective surface tension must still be established. However, there are also other mathematical problems to be solved, resulting from the unicity of orthogonal 3D meshes, from surfaces that present singularities and so on.

Many evolutions can be solved using equations for continuous media, sometimes analytically but more often numerically. Interfaces also present a new important problem, as it is impossible to know beforehand, what shape the surfaces will take, nor their equations.

Equations for systems containing interfaces are generally numerically solved. The question of following the movement and capillary interfaces under strain has been the subject of much research and digital codes have been developed for this purpose (e.g. *gerris* and *basilisk*, Popinet 2018). For problems with thick interfaces, we have opted for solutions in volume, by refining the meshing of interfacial zones considered as continuous media[6]. In some cases, we had to artificially thicken the interfacial layers to facilitate the solution and thus introduce phase field methods at the appropriate scale to avoid steep interface jumps.

The objective of our approach was to present the concept of an interface with its thermodynamic and mechanical modeling. We hope to have contributed to this in this short chapter, even as we acknowledge that a much more exhaustive study must be carried out. Updates appear more or less regularly in specialized journals, which also cover the topic of membranes. Fibers and ligaments have not been discussed in this chapter, nor did we examine radiation and the effects of electromagnetic fields.

For more details on the subjects discussed here, the reader will find it useful to consult the additional references provided below.

---

6 This is difficult to implement for capillary surfaces that are too thin.

1.6. References

Akamatsu, S. and Faivre, G. (2000). Traveling waves, two-phase fingers, and eutectic colonies in thin sample directional solidification of a ternary eutectic alloy. *Phys. Rev. E*, 61, 3757–3769.

Akamatsu, S., Bottin-Rousseau, S., Faivre, G., Ghosh, S., Plapp, M. (2015). Lamellar eutectic growth with anisotropic interphase boundaries. *IOP Conference Series: Materials Science and Engineering*, IOP Publishing, 84, 012083.

Bertrand, G. and Prud'homme, R. (1977). Possibilities of surface reaction coupling with transport phenomena. *Int. J. Quant. Chem.*, 12(2), 159168.

Clavin, P. and Joulin, G. (1983). Premixed flames in large and high intensity turbulent flow. *Le J. de Phys Lett.*, 44(1), L1–L12.

Defay, R. (1949). Essai d'une thermodynamique des couches minces. *J. Chim. Phys.*, 46, 375–387.

Defay, R. and Prigogine, I. (1946). Théorie thermodynamique de la tension superficielle dynamique. *J. Chim. Phys.*, 43, 217–234.

Defay, R. and Sanfeld, A. (1973). Etats transitoires de tension superficielle nulle. *J. de chimie physique*, 70, 895–899.

Defay, R., Sanfeld, A., Steinchen, A. (1972). La couche interfaciale traitée comme une collection de plans moléculaires II. Comparaison avec le modèle continu. *J. Chim. Phys.*, 69, 1380–1385.

Duraiswami, R. and Prosperetti, A. (1992). Orthogonal mapping in two dimensions. *J. Computat. Phys.*, 98, 254–268.

Gatignol, R. and Prud'homme, R. (2001). Mechanical and thermodynamical modeling of fluid interfaces. *Series on Advances in Mathematics for Applied Sciences*, 58, World Scientific.

Germain, P. (1987). *Mécanique*. Ellipses, Ecole Polytechnique, Palaiseau.

Huffenus, J.-P. (1969). Calculs d'écoulements en fluide réel. *La Houille Blanche*, (6), 597–602.

Jamet, N. (1998). Modélisation des discontinuités dans les écoulements diphasiques compressibles. PhD Thesis, Université Pierre et Marie Curie, Paris.

Jamet, N., Prud'homme, R., Gottesdiener, L. (1997). Coefficients phénoménologiques au travers de discontinuités en milieu diphasiques. *13ème congrès français de mécanique*, Poitiers-Futuroscope.

Jaumotte, A.L. (ed.) (1971). *Chocs et ondes de choc*. Masson, Paris.

Karma, A. and Sarkissian, A. (1996). Morphological instabilities of lamellar eutectics. *Met. Trans. A*, 27, 635.

Kervella, Y., Khojasteh Pour Fard, I., Le Hir, P., Renault, E., Lemoine, J.-P. (2012). Modélisation hydrodynamique tridimensionnelle en coordonnées curvilignes non-orthogonales de l'estuaire de la Seine. *XIIèmes Journées Nationales Génie Côtier – Génie Civil*, Cherbourg.

Marmottant, P.H. and Villermaux, E. (2004). On spray formation. *J. Fluid Mech.*, 498, 73–111.

Petitjeans, P., Kurowski, P., Fernandez, J., Hoyos, M. (1999). Instabilités à "l'interface" entre deux fluides miscibles. Émergence d'une tension de surface effective. *J. Chim. Phys.* 96, 991–996.

Pojman, J.A., Whitmore, C., Turco Liveri, M.L., Lombardo, R., Marszalek, J., Parker, R., Zoltowski, B. (2006). Evidence for the existence of an effective interfacial tension between miscible fluids: Isobutyric acid–water and 1-butanol-water in a spinning-drop tensiometer. *Langmuir*, 22, 2569–2577.

Popinet, S. (2018). Numerical models of surface tension. *Annual Review of Fluid Mechanics*, 50, 49–75.

Prigogine, I. and Defay, R. (1949). Tension superficielle dynamique des solutions régulières. *J. Chim. Phys.*, 46, 367–372.

Prud'homme, R. (1988). *Fluides hétérogènes et réactifs : écoulements et transferts*. Springer-Verlag, Berlin.

Prud'homme, R. (2010). *Flows of Reactive Fluids*. Springer, New York, Dordrecht, Heidelberg, London.

Prud'homme, R. (2012). *Flows and Chemical Reactions*. ISTE Ltd, London and John Wiley & Sons, New York.

Renaud-Assemat, I. (2011). Développement d'une méthode lagrangienne de simulation d'écoulements turbulents à phases séparées. PhD Thesis, Inst. Nat. Polytech. de Toulouse, Toulouse.

Rocard, Y. (1933). Equations hydrodynamiques avec termes capillaires : théorie de la tension superficielle. *Le journal de physique et le radium*, VII(IV), 10.

Roussel, J. (2016). Cours de mécanique des fluides – tension superficielle, Cours ENSCR [Online]. Available at: http://femto-physique.fr/mecanique_des_fluides/mecaflu_C4.php.

Ryskin, G and Leal, L.G. (1983). Orthogonal mapping. *Journal of Computational Physics*, 50, 71–100.

Ryskin, G. and Leal, L.G. (1984). Numerical solution of free boundary problem in fluid mechanics, part 1: The finite difference technique. *J. Fluid Mech.*, 148, 1–18.

Seppecher, P. (1987). Etude d'une modélisation des zones capillaires fluides : interfaces et zones de contact. PhD Thesis, Université Paris VI, Paris.

Slattery, J.C. (1990). *Interfacial Transport Phenomena*. Springer, Berlin.

Smolderen, J. (1971). Structure des chocs et théorie cinétique des gaz. In *Chocs et ondes de choc*, Jaumotte, A.L. (ed.). Masson, Paris.

Further references

Blanco, A. (1992). Génération de maillages orthogonaux. Mémoire de D.E.A. Inst. Polytechnique de Toulouse, Toulouse.

Blanco, A. (1995). Quelques aspects de l'écoulement d'un fluide visqueux autour d'une bulle déformable : une analyse par simulation directe. PhD Thesis, Inst. Polytechnique de Toulouse, Toulouse.

Blanco, A. and Magnaudet J. (1995). The structure of the axisymmetric high-Reynolds number flow around an ellipsoidal bubble of fixed shape. *Phys. Fluid*, 7(6).

Caramero, R. (2009). Génération de maillages, Département de génie mécanique. Ecole Polytechnique de Montréal, Montreal.

Del Piero, G. (2009). On the method of virtual power in continuum mechanics. *Journal of Mechanics of Materials and Structures*, 4(2).

Fleury, P. and Mathieu, J.P. (1965). *Mécanique physique*. Eyrolles, Paris.

de Gennes, P.-G., Brochard Wyart, F., Quéré, D. (2002). *Gouttes, bulles, perles et ondes*. Belin, Paris.

Groupement de Recherche (n/a). Résumés des travaux des colloques annuels du GDR MFA. Available at: https://www.gdr-mfa.fr/.

Jamet, N., Prud'homme, R., Gottesdiener, L. (1999). Modeling of discontinuities in two-phase compressible flows. Unpublished paper, Research Gate, February 2015.

Langevin, D. (1998). La coalescence. *Bulletin de la SFP*, 115.

Legendre, M., Petitjeans, P., Kurowski, P. (2001). Instabilités à l'interface entre fluides miscibles par forçage oscillant horizontal. *C. R. Acad. Sci. Paris*, IV(329), 1.

Léger, L. (1989). L'étalement d'une goutte sur un solide : rôle des forces à longue portée. *Images de la physique – CNRS*, 74–79.

Prud'homme, R. (2014). *Flows and Chemical Reactions in Heterogeneous Mixtures*. ISTE Ltd, London, and John Wiley & Sons, New York.

Quéré, D. (2005). Les surfaces superhydrophobes. *Images de la physique – CNRS*, 239–244.

2

Simulations of Turbulent Two-Phase Flows with Phase Change Using a Multifield Approach Combined with LES[1]

Safety issues in nuclear power plants involve complex turbulent bubbly flows. To predict the behavior of these flows, the two-fluid approach is often used. Nevertheless, this model has been developed for the simulation of small spherical bubbles, considered as a dispersed field. To deal with bubbles with a large range of sizes, a multifield approach based on this two-fluid model has been proposed. A special treatment, called the Large Bubble Model (LBMo), has been implemented and coupled to the dispersed model. However, only laminar and isothermal flows were considered in previous papers. Here, large-eddy simulations (LES) are investigated to model turbulence effects. For this purpose, the two-fluid model equations are filtered to highlight the specific subgrid terms. Then, an *a priori* LES study using filtered direct numerical simulation (DNS) results is detailed. This analysis allows us to classify these terms according to their relative weight and then concentrate the modeling efforts on the predominant ones. Five different turbulence models are compared. These results are finally used to perform true LES on a turbulent two-phase flow. Moreover, in order to tackle non-isothermal flows occurring in industrial studies, a new heat transfer model is implemented and validated to deal with phase change at large interfaces using the Large Bubble Model.

Chapter written by Solène GOUÉNARD, Stéphane VINCENT and Stéphane MIMOUNI.

1 The results presented in this chapter have been obtained with the CMFD tool NEPTUNE_CFD, developed in the framework of the MULTIPATH project, financially supported by the CEA (Commissariat l'Énergie Atomique et aux Energies Alternatives), EDF (Électricité de France), IRSN (Institut de Radioprotection et de Sûreté Nucléaire) and AREVA NP, in collaboration with the MSME laboratory (Université Paris-Est Marne-la-Vallée, Champs sur Marne, France). The authors are grateful for access to the computational facilities of the French CINES (National computing center for higher education) and CCRT (National computing center of CEA) under project number x20172b6115.

List of symbols

$\Delta\tau$	s	Characteristic time scale
ξ	–	Convergence order
Δx	m	Cube root of the cell volume
ρ	$kg.m^{-3}$	Density
μ	$kg.m^{-1}.s^{-1}$	Dynamic viscosity
$\widehat{\Delta}$	m	Filter width
κ	m^{-1}	Interface curvature
Γ	$kg.m^{-3}.s^{-1}$	Interfacial mass transfer
ϵ	–	Relative error
λ	$W.m^{-1}.K^{-1}$	Thermal conductivity
χ	$m^2.s^{-1}$	Thermal diffusivity
σ	$N.m^{-1}$	Surface tension coefficient
Ω	m^3	Volume of a cell
ε^*	–	Volume fraction

Greek letters

$\overline{}$	LES filter
i, j	Space directions
I	Cell index
k	Field index
v	Vapor phase
l	Liquid phase
Int	Interface

Subscripts and superscripts

C_p	$J.m^{-1}.K^{-1}$	Specific heat capacity
d_p	m	Characteristic length scale
g	$m.s^{-2}$	Gravity acceleration
G	–	LES filter
h	–	Heaviside function
H	$J.kg^{-1}$	Enthalpy
I_d	–	Identity matrix
L	$J.kg^{-1}$	Latent heat
n	–	Unit interface normal vector
P	Pa	Pressure
Q	$W.m^2$	Conductive thermal flux
S	s^{-1}	Viscous stress tensor
T	K	Temperature
u	$m.s^{-1}$	Velocity

Roman letters

ADM	Approximate Deconvolution Model
CSF	Continuum Surface Force
CMFD	Computational Multifluid Dynamics
CPU	Central Processing Unit
DNS	Direct Numerical Simulation
LBMo	Large Bubble Model
LES	Large-Eddy Simulation
VOF	Volume of Fluid
WALE	Wall-Adapting Local Eddy-Viscosity

Acronyms

2.1. Introduction

Turbulent non-isothermal flow studies with Computational Multifluid Dynamics (CMFD) tools require the ability to deal with inclusions of different sizes and shapes, turbulence effects and phase change. To deal with the first issue, two approaches can be considered. Bubbly flows occurring in nuclear power plants or in conventional heat exchangers are often modeled with an Eulerian dispersed description within the two-fluid model by Ishii (1975). In this model, the bubbles are small enough to be considered spherical. Thus, interfacial forces such as drag force, lift force, wall lubrication, virtual mass and turbulent dispersion force are implemented (e.g. Mimouni *et al.* 2011). However, this approach is not suitable for the simulation of large and deformable bubbles. In the second type of approaches, inclusions are located using interface tracking methods such as front-tracking (e.g. Unverdi and Tryggvason 1992), level-set (e.g. Sussman *et al.* 1994) or volume of fluid (VOF) (e.g. Hirt and Nichols 1981). Nevertheless, to simulate flows containing inclusions of different sizes, the grid has to be refined according to the smallest interfacial scale. In industrial configurations, these methods are not sustainable in terms of central processing unit (CPU) consumption.

More recently, new categories of methods have been developed by combining the two previous approaches for the simulation of flows containing inclusions with a large range of sizes. In the first type of approach, the larger scales are simulated using the interface tracking methods described above, and the badly resolved structures are considered as dispersed particles followed by a Lagrangian method. This has been implemented by several research groups with the level-set method (e.g. Capecelatro *et al.* 2010; Herrmann 2010) and the VOF method (e.g. Tomar *et al.* 2010; Ling *et al.* 2015). A second type of approach, called the multifield approach, was introduced a long time ago by Anderson and Jackson (1967) and Drew and Lahey (1979). Contrary to the first method, the multifield approach is based on a two-fluid model initially used to simulate complex flows with a continuous phase and spherical inclusions modeled through a dispersed approach. It has been implemented by different research groups with a Lagrangian description of the small spherical inclusions (e.g. Zuzio *et al.* 2013) and with Eulerian methods (e.g. Vallet and Borghi 1999; Lebas *et al.* 2009). In this chapter, the second type of approach based on an Eulerian description of the smallest inclusions is used. The two-fluid model is then adapted for the simulations of large and deformable inclusions by considering a second continuous field in the balance equations. The large inclusions are considered as an interface between two continuous fields, whereas the small spherical inclusions are treated as a dispersed field. Specific treatments are required to simulate these interfaces, which are referred to as the Large Bubble Model (LBMo). This model has been extensively validated on laminar isothermal flows in previous papers (e.g. Denèfle *et al.* 2015; Fleau *et al.* 2015). The criteria to switch from the large interfaces treated with the LBMo to the dispersed field and the opposite are not given in this chapter since we are focusing on the large interfaces. Readers are invited to refer to Mimouni *et al.* (2017) for details. However, industrial studies require the ability to deal with both turbulence and phase change effects.

Thus, the second main issue in complex turbulent flows, considered in this chapter, is turbulence modeling in the presence of interfaces. Within the multifield approach, large interfaces are described in a deterministic way through the volume fraction, updated in the mass balance equation. Therefore, to simulate turbulent flows, the large-eddy simulation (LES) seems more suitable since it is also based on a deterministic description of the turbulence effects. The simulation of turbulent two-phase flows containing large interfaces has never been performed with two-fluid models. However, a great deal of work has been conducted on single-fluid models to analyze the subgrid terms, quantify them and compare different turbulence models (e.g. Labourasse *et al.* 2007; Vincent *et al.* 2008; Larocque *et al.* 2010). Thus, in this chapter, it is proposed to follow the same methodology applied to single-fluid models, to the two-fluid model with the LBMo.

Finally, the third issue consists of the treatment of phase change at large interfaces using the LBMo. Different phase change models are available in literature. The most widely used is the model proposed by Lee (1980), based on an empirical expression to quantify the interfacial heat and mass transfer term. Nevertheless, a coefficient has to be fixed to bring the temperature of the cells containing the interface close to the saturation temperature. A large range of values has been obtained from 0.1 s^{-1} (e.g. Wu *et al.* 2007; Schepper *et al.* 2009; Alizadehdakhel *et al.* 2010) to 100 s^{-1} (e.g. Yang *et al.* 2008; Fang *et al.* 2010). It has also been shown that too large values could induce numerical instabilities with oscillations of the temperature around the saturation temperature. To avoid the issue of fixing an empirical coefficient, the development of a formulation based on theoretical considerations is necessary. Thus, to estimate the heat flux jump at the interface, Fourier's law is applied. This model and other similar approaches have been used by several research groups (e.g. Mao 2009; Nichita and Thome 2010; Kunkelmann *et al.* 2012; Ganapathy *et al.* 2013; Sun *et al.* 2014) with the VOF or level-set method. Nevertheless, all of these models are applied to interfaces contained in one grid cell. In the framework of the multifield approach with the two-fluid model, the large interfaces are diffused over a few cells. The expression of the heat flux has to be adapted to interfaces with a non-zero thickness. Thus, the development, implementation and validation of a new heat transfer model based on the methodology proposed by Brackbill *et al.* (1992), for the surface tension force, is detailed.

After presenting the computational model, an *a priori* LES study will be performed to evaluate the order of magnitude of the subgrid terms appearing in the filtered two-fluid model equations, and to compare the different turbulence models usually applied to single-phase flows. For this purpose, the phase inversion benchmark (e.g. Vincent *et al.* 2016) will be simulated. Then, the more suitable turbulence model will be assessed on a horizontal pipe of water and air. The objective of this test case is to highlight the ability of the code to predict flow regimes. This last point is crucial in nuclear power plants to improve the performance and life span of pipes. Finally, non-isothermal flows will be considered. A new phase change model for large interfaces will be implemented and validated.

2.2. Computational model

In this section, the two-fluid model is presented in the context of large and deformable interfaces. The balance equations are solved for each field. Thus, in the framework of the multifield approach, for a two-phase flow with water and steam inclusions of different sizes, three fields are considered: a continuous liquid field, a continuous gas field (contained in the large deformable steam inclusions) and a dispersed gas field (the small spherical steam inclusions). Then, the LBMo, implemented for the simulation of interfaces between two continuous fields, is detailed.

2.2.1. *Two-fluid model*

The code NEPTUNE_CFD is based on an Eulerian approach with a finite volume discretization. The flow motion is followed by using the two-fluid model by Ishii (1975) extended to n-phase. In this model, the density, the viscosity and the local velocity are defined for each field in each cell. The following governing equations are solved for each field k:

– The mass balance equation:

$$\partial_t \varepsilon_k^* \rho_k + \partial_{x_i}\left(\varepsilon_k^* \rho_k u_{i,k}\right) = \Gamma_k \qquad [2.1]$$

where ε_k^* is the volume fraction of field k, ρ_k is its density, $u_{i,k}$ is the ith component of the velocity of field k, and Γ is the mass transfer term.

– The momentum balance equation in each space direction i:

$$\begin{aligned} \partial_t(\varepsilon_k^* \rho_k u_{i,k}) + \partial_{x_j}\left(\varepsilon_k^* \rho_k u_{i,k} u_{j,k}\right) &= \mu_k \partial_{x_j}\left(\varepsilon_k^* S_{ij,k}\right) - \varepsilon_k^* \partial_{x_j} P \\ &\quad + \varepsilon_k^* \rho_k g_i + F_{i,k} \end{aligned} \qquad [2.2]$$

where μ_k denotes the viscosity of field k, S denotes the viscous stress tensor, P denotes the pressure, g denotes the gravitational constant, and $F_{i,k}$ denotes the specific source terms depending on the field (continuous or dispersed).

– The energy balance equation:

$$\begin{aligned} \partial_t(\varepsilon_k^* \rho_k H_k) + \partial_{x_j}\left(\varepsilon_k^* \rho_k H_k u_{j,k}\right) &= - \partial_{x_j}(\varepsilon_k^* Q_{j,k}) + \mu_k \partial_{x_j}\left(\varepsilon_k^* \mu_k S_{ij,k} u_{j,k}\right) \\ &\quad + \varepsilon_k^* \partial_t P + \varepsilon_k^* \rho_k g_i u_{i,k} + E_k^{Int} \end{aligned} \qquad [2.3]$$

where H_k is the total enthalpy, $\boldsymbol{Q}_k = -\lambda_k \nabla T_k$, is the conductive thermal flux, λ_k is the thermal conductivity, T_k is the temperature, and E_k^{Int} are some other interfacial energy transfers. The energy balance equation is only solved when non-isothermal flows are considered, which will only be discussed in the final section.

In the code NEPTUNE_CFD, the assumption of a common pressure for all fields is made.

The simulations are performed on fixed grids with collocated variables. The data structure is totally face-based, which allows the use of arbitrary-shaped cells including non-conforming meshes (e.g. Guelfi *et al.* 2007). The well-known semi-implicit method for pressure-linked equations (SIMPLE) solver is used in the CMFD tool (e.g. Patankar and Spalding 1972). An iterative coupling of the equations is implemented to ensure mass and energy conservation.

2.2.2. *Large Bubble Model*

The LBMo (e.g. Denèfle *et al.* 2015; Fleau *et al.* 2015) has been developed to simulate large interfaces with a two-fluid model. This model consists of the three elements necessary for an accurate simulation of large and deformable inclusions.

2.2.2.1. *Surface tension*

The surface tension model used in the NEPTUNE_CFD code for the simulation of the large interfaces corresponds to the volumetric formulation (e.g. Bartosiewicz *et al.* 2008; Denèfle *et al.* 2015; Fleau *et al.* 2015) of the continuum surface force (CSF) model proposed by Brackbill *et al.* (1992):

$$\boldsymbol{F}_{\boldsymbol{CSF}} = \varepsilon_k^* \sigma \kappa \nabla \varepsilon_k^* \Omega \qquad [2.4]$$

where σ is the surface tension coefficient, Ω is the cell volume, and κ is the local curvature:

$$\kappa = -\nabla.\left(\frac{\nabla \varepsilon_k^*}{\left\|\nabla \varepsilon_k^*\right\|}\right) \qquad [2.5]$$

This surface tension force appears in the momentum balance equation applied to continuous fields through the term $F_{i,k}$.

2.2.2.2. *Drag force*

In the two-fluid model, the drag force is crucial. Contrary to the single-fluid approach, two different velocities are defined, one for each field. The role of the drag force law is to couple these two velocities at the interface. Thus, a new drag force expression has been developed to deal with large interfaces within the multifield approach (e.g. Fleau *et al.* 2015):

$$\begin{aligned} &\varepsilon_2^* < 0.3 : \mathbf{F_{bubble}} = \varepsilon_1^* \varepsilon_2^* \frac{18\mu_1}{\varepsilon_1^* d_p^2} (\boldsymbol{u_1} - \boldsymbol{u_2}) \\ &\varepsilon_2^* > 0.7 : \mathbf{F_{droplet}} = \varepsilon_1^* \varepsilon_2^* \frac{18\mu_2}{\varepsilon_2^* d_p^2} (\boldsymbol{u_1} - \boldsymbol{u_2}) \\ &0.3 \leq \varepsilon_2^* \leq 0.7 : \mathbf{F_{mix}} = \frac{0.7 - \varepsilon_2^*}{0.7 - 0.3} \mathbf{F_{bubble}} + \frac{\varepsilon_2^* - 0.3}{0.7 - 0.3} \mathbf{F_{droplet}} \end{aligned} \qquad [2.6]$$

The characteristic length scale d_p is equal to $\sqrt{\frac{\mu_2}{\mu_1} \frac{\varepsilon_k^*}{\|\nabla \varepsilon_k^*\|}}$ close to the large interfaces, for $0.1 \leq \varepsilon_1^* \varepsilon_2^* \leq 0.25$, and to the diameter of the dispersed bubbles/droplets far from them, for $\varepsilon_1^* \varepsilon_2^* \leq 0.02$. Between these two regions, a linear extrapolation of these two values is defined.

This drag force appears in the momentum balance equation applied to continuous fields through the term $F_{i,k}$.

2.2.2.3. *Interface sharpening*

To control the interface thickness, the interface sharpening equation proposed by Olsson and Kreiss (2005), and adapted to the two-fluid formulation, is solved for each continuous field (Fleau *et al.* 2016):

$$\partial_\tau \varepsilon_k^* + \nabla . \left(\varepsilon_k^* \left(1 - \varepsilon_k^* \right) \boldsymbol{n} \right) = K \Delta \varepsilon_k^* \qquad [2.7]$$

where $\boldsymbol{n} = \frac{\nabla \varepsilon_k^*}{\|\nabla \varepsilon_k^*\|}$ is the interface normal vector. The values of the parameters $\Delta\tau = \frac{\Delta x}{32}$ and $K = \frac{\Delta x}{2}$ (Δx denotes the cell size) are chosen to obtain a final interface thickness always equal to five cells whatever the initial interface diffusion (e.g. Štrubelj 2009; Denèfle *et al.* 2015).

The LBMo has been widely validated on various isothermal test cases (e.g. Fleau *et al.* 2015; Duponcheel *et al.* 2017; Fleau *et al.* 2016; Vincent *et al.* 2016; Mimouni *et al.* 2017). Reasonable agreements have been obtained with analytical and experimental data on free surfaces and bubble test cases in laminar and turbulent conditions.

2.3. Filtered two-fluid equations

As previously done in Labourasse *et al.* (2007) and Vincent *et al.* (2008) with the single-fluid model and Lakehal (2004) with the two-fluid model including a dispersed field, the LES filter is applied to the two-fluid model equations for two continuous fields within the LBMo. In this section, the flows are considered isothermal. Thus, the filtered energy balance equation is not detailed. No dispersed fields are taken into account.

– The filtered mass balance equation:

$$\rho_k \partial_t \overline{\varepsilon_k^*} + \rho_k \nabla . \left(\overline{\varepsilon_k^*} \, \overline{u_{i,k}} \right) + \tau_{interf} = 0 \qquad [2.8]$$

where $\overline{\varepsilon_k^*}$ is the filtered volume fraction of field k and τ_{interf} is a subgrid term related to the relationship between the filtered velocity $\overline{u_{i,k}}$ and the interface topology (see Table 2.1).

– The filtered momentum balance equation:

$$\begin{aligned}\rho_k \partial_t \left(\overline{\varepsilon_k^*} \, \overline{u_{i,k}} \right) + \tau_{time} + \rho_k \nabla . \left(\overline{\varepsilon_k^*} \, \overline{u_{i,k}} \, \overline{u_{j,k}} \right) + \tau_{conv} \\ = \mu_k \nabla . \left(\overline{\varepsilon_k^*} \, \overline{S_k} \right) + \tau_{diff} - \overline{\varepsilon_k^*} \nabla \overline{P} - \tau_{pressure} \\ + \overline{\varepsilon_k^*} \rho_k g_i + \widehat{F_{CSF}} + \tau_{superf} + \widehat{F_{drag}} + \tau_{drag}\end{aligned} \qquad [2.9]$$

Subgrid terms	Expressions
τ_{interf}	$\rho_k \left(\nabla . \left(\overline{\varepsilon_k^* u_{i,k}} \right) - \nabla . \left(\overline{\varepsilon_k^*} \, \overline{u_{i,k}} \right) \right)$
τ_{time}	$\rho_k \left(\partial_t \left(\overline{\varepsilon_k^* u_{i,k}} \right) - \partial_t \left(\overline{\varepsilon_k^*} \, \overline{u_{i,k}} \right) \right)$
τ_{conv}	$\rho_k \left(\nabla . \left(\overline{\varepsilon_k^* u_{i,k} u_{j,k}} \right) - \nabla . \left(\overline{\varepsilon_k^*} \, \overline{u_{i,k}} \, \overline{u_{j,k}} \right) \right)$
τ_{diff}	$\mu_k \left(\nabla . \left(\overline{\varepsilon_k^* S_k} \right) - \nabla . \left(\overline{\varepsilon_k^*} \, \overline{S_k} \right) \right)$
$\tau_{pressure}$	$\overline{\alpha_k \nabla P} - \overline{\alpha_k} \nabla \overline{P}$
τ_{superf}	$\sigma \left(\overline{\varepsilon_k^* \kappa \nabla \alpha_k} - \overline{\varepsilon_k^*} \, \hat{\kappa} \, \overline{\nabla \alpha_k} \right)$

Table 2.1. *Expression of the subgrid terms appearing in the filtered two-fluid equations, with* $\hat{\kappa} = -\nabla . \left(\frac{\nabla \overline{\varepsilon_k^*}}{\left\| \nabla \overline{\varepsilon_k^*} \right\|} \right)$ *being the filtered local curvature*

where τ_{time}, τ_{conv}, τ_{diff} are, respectively, the time, convective and diffusive subgrid terms, and $\tau_{pressure}$, τ_{superf} and τ_{drag} are the three specific subgrid terms of the two-fluid model applied to two continuous fields (see Table 2.1 for the expressions). τ_{drag} has the following expression:

$$\tau_{drag} = \overline{F_{drag}} - \widehat{F_{drag}} \quad [2.10]$$

with $\overline{F_{drag}}$ being defined as:

$$\begin{aligned}
&\overline{\varepsilon_2^*} < 0.3 : \overline{\mathbf{F_{bubble}}} = \overline{\varepsilon_1^* \varepsilon_2^* \frac{18\mu_1}{\varepsilon_1^* d_p^2}(\boldsymbol{u_1} - \boldsymbol{u_2})} \\
&\overline{\varepsilon_2^*} > 0.7 : \overline{\mathbf{F_{droplet}}} = \overline{\varepsilon_1^* \varepsilon_2^* \frac{18\mu_2}{\varepsilon_2^* d_p^2}(\boldsymbol{u_1} - \boldsymbol{u_2})} \\
&0.3 \le \overline{\varepsilon_2^*} \le 0.7 : \overline{\mathbf{F_{mix}}} = \overline{\frac{0.7\text{-}\varepsilon_2^*}{0.7\text{-}0.3}\mathbf{F_{bubble}} + \frac{\varepsilon_2^*\text{-}0.3}{0.7\text{-}0.3}\mathbf{F_{droplet}}}
\end{aligned} \quad [2.11]$$

and $\widehat{F_{drag}}$:

$$\begin{aligned}
&\overline{\varepsilon_2^*} < 0.3 : \widehat{\mathbf{F_{bubble}}} = \overline{\varepsilon_1^*}\ \overline{\varepsilon_2^*} \frac{18\mu_1}{\overline{\varepsilon_1^*} \widehat{d_p}^2} (\overline{\boldsymbol{u_1}} - \overline{\boldsymbol{u_2}}) \\
&\overline{\varepsilon_2^*} > 0.7 : \widehat{\mathbf{F_{droplet}}} = \overline{\varepsilon_1^*}\ \overline{\varepsilon_2^*} \frac{18\mu_2}{\overline{\varepsilon_2^*} \widehat{d_p}^2} (\overline{\boldsymbol{u_1}} - \overline{\boldsymbol{u_2}}) \\
&0.3 \le \overline{\varepsilon_2^*} \le 0.7 : \widehat{\mathbf{F_{mix}}} = \frac{0.7\text{-}\overline{\varepsilon_2^*}}{0.7\text{-}0.3} \widehat{\mathbf{F_{bubble}}} + \frac{\overline{\varepsilon_2^*}\text{-}0.3}{0.7\text{-}0.3} \widehat{\mathbf{F_{droplet}}}
\end{aligned} \quad [2.12]$$

$\widehat{F_{CSF}}$ can be written in the same way as $\widehat{F_{drag}}$ by applying the LES filter at each quantity. We can note that, compared to the one-fluid model, specific subgrid terms appear in the filtered two-fluid model equation, which are $\tau_{pressure}$, τ_{drag}. Thus, seven subgrid terms have to be modeled (see Table 2.1).

2.4. *A priori* LES study

After presenting the filtered equations and the specific subgrid terms appearing, the phase inversion benchmark is simulated with the LBMo to perform an *a priori* LES study. Only continuous fields are considered for this simulation. This test case is turbulent and isothermal. A validation of the results can be found in Vincent *et al.* (2016). In this analysis, the value of the volume fractions, pressure and velocities are filtered. This allows us to classify the subgrid terms according to their weight. Then, turbulence models are applied and compared.

2.4.1. *Presentation of the test case*

In this test case, an oil drop with a cubic shape (size of L/2) is initially placed in a cubic box (size L = 0.1 m) containing liquid water (see Figure 2.1). The evolution of the system is driven by the gravity forces in the z direction (g = 9.81 m.s^{-2}). At the end of the simulation, the oil phase is supposed to be located in the top part of the box with the liquid water beneath. The fluid properties are given in Table 2.2. The interfacial tension coefficient between oil and water is equal to 0.045 N.m^{-1}.

The test case has been simulated with three different mesh refinements: 128^3 cells, 256^3 cells and 512^3 cells. The time steps are kept constant and are, respectively, equal to 0.8 ms, 0.2 ms and 0.05 ms, such that the CFL number is constant and always smaller than 0.9. The simulations have been performed with 144 cores for the first grid and 1152 for the two others during, respectively, 7 hours, 47 hours and 2 months and a half to reach 13 physical seconds.

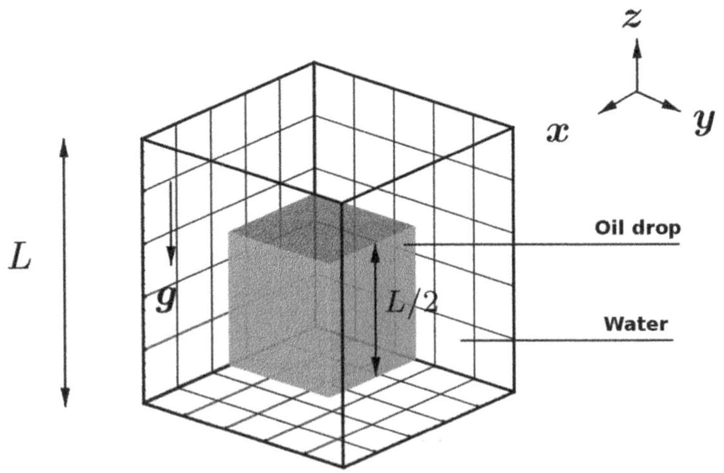

Figure 2.1. *Initial conditions of the phase inversion benchmark. For a color version of this figure, see www.iste.co.uk/prudhomme/fluid1.zip*

Phase	ρ (kg.m^{-3})	μ (kg.m^{-1}.s^{-1})
Oil	900	0.1
Water	1000	$1 \cdot 10^{-3}$

Table 2.2. *Physical properties of the two phases in the phase inversion benchmark*

2.4.2. *Order of magnitude of the subgrid terms*

The subgrid terms appearing in the filtered two-fluid equations are compared in terms of order of magnitude to find the predominant and negligible ones.

This first step is important, since the larger the contribution of this subgrid term is, the more the modeling errors will impact the simulation. For this purpose, a uniform box filter is applied to the simulation results extracted at the peak of enstrophy (maximum value of the velocity curl) for the three grids. Only the first neighborhood of each cell (filter size of 2) is considered. The subgrid term τ_{time} is not represented in this study, since the analysis is only proposed for one time. Each subgrid term in the momentum balance equation is normalized by the convection resolved term. The interfacial subgrid term τ_{interf} appearing in the mass balance equation is normalized by its corresponding resolved part.

The summation of the subgrid term and the convection (respectively interfacial) resolved term is made in the whole domain, only where the phase is present, as illustrated here with the convection subgrid term:

$$\overline{\tau_{conv}} = \frac{\sum_{\overline{\varepsilon_k^*}^I > 1\cdot 10^{-10}} \left|\overline{\tau_{conv}}^I\right|}{\sum_{\overline{\varepsilon_k^*}^I > 1\cdot 10^{-10}} \left|\rho_k \nabla . \left(\overline{\varepsilon_k^*}^I \, \overline{u_{i,k}}^I \, \overline{u_{j,k}}^I\right)\right|} \quad [2.13]$$

The results are presented in Figure 2.2. Table 2.3 proposes a classification of the subgrid terms according to the phase for the most refined grid. As documented in Labourasse *et al.* (2007), Vincent *et al.* (2008) and Larocque *et al.* (2010), the subgrid term τ_{superf} is largely predominant for the two phases. Furthermore, a small contribution of the diffusion subgrid term τ_{diff} is observed for the two phases. Nevertheless, for the oil phase, contrary to the single-fluid formulation, this contribution is not negligible and very close to those of τ_{conv} and τ_{interf}. The results obtained with the single-fluid formulation seem to be closer to those obtained here for the water phase.

As expected, the magnitude of each subgrid term decreases when the mesh is refined, except τ_{superf} for the oil phase. This increase has been previously observed with single-fluid formulations (e.g. Vincent *et al.* 2016). In the two-fluid model formulation, this evolution is the result of two opposite effects:

– The first effect is related to the ratio between the numerical and the laminar viscosities of each phase. When the mesh is refined, the velocities increase, since the numerical viscosity decreases. The water phase is more sensitive to a variation of the numerical viscosity, since its laminar viscosity is smaller. Thus, the velocities in the water phase increase more than in the oil phase, which has a larger laminar viscosity. The resolved convection term then increases more in the water phase than in the oil phase.

– The second effect is the number of small inclusions, which increases when the mesh is refined and induces an increase of the subgrid term.

The evolution of the subgrid term contribution then depends on which effect is predominant. For the oil phase, it seems that the increase of small inclusions drive the evolution of the subgrid term, whereas for the water phase, the decrease of the numerical viscosity is predominant.

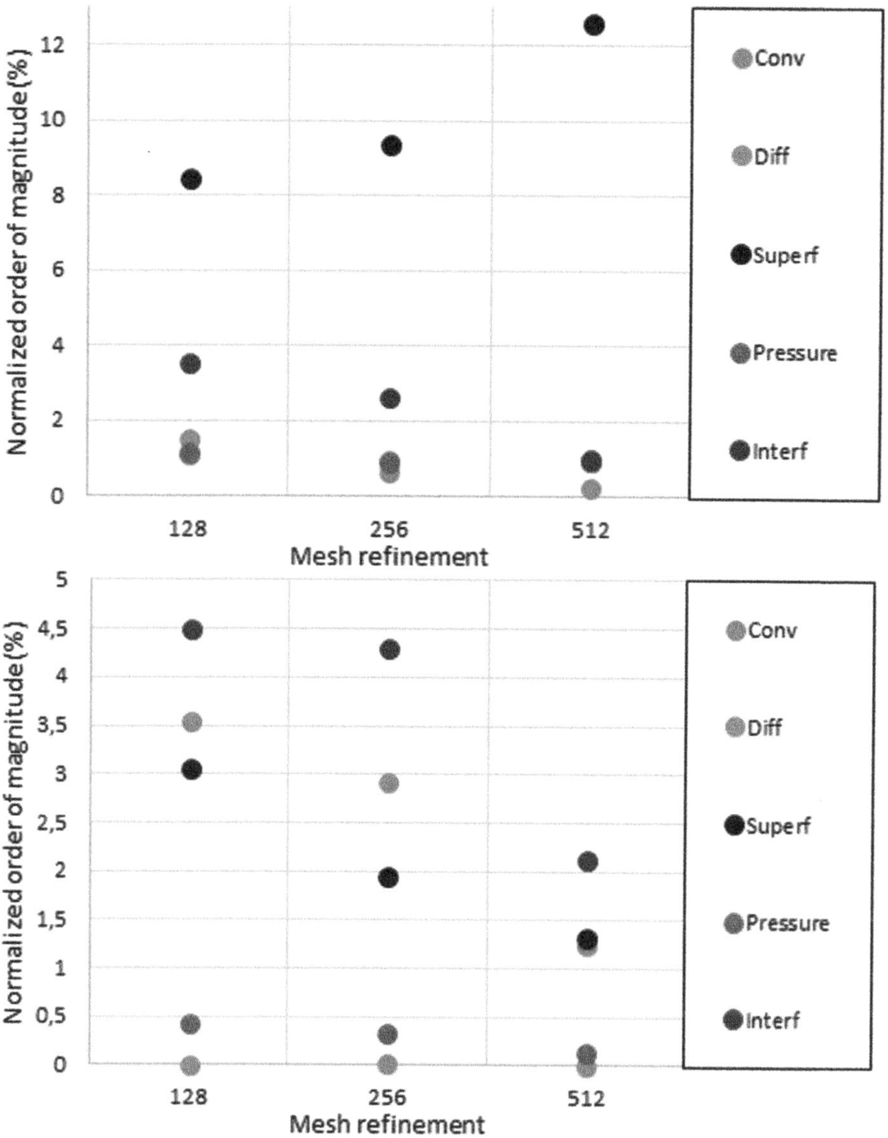

Figure 2.2. *Order of magnitude of the normalized subgrid terms, top: for the oil phase and bottom: for the water phase, only the z component is displayed. For a color version of this figure, see www.iste.co.uk/prudhomme/fluid1.zip*

	Oil	Water
1	τ_{superf}	τ_{superf}
2	$\tau_{pressure}$	τ_{conv}
3	τ_{diff}	$\tau_{pressure}$
4	τ_{conv}	τ_{interf}
5	τ_{interf}	τ_{diff}

Table 2.3. *Classification of the subgrid terms according to their relative contribution for the 512^3 grid with the box filter*

Finally, the contribution of the drag subgrid term τ_{drag} is not mentioned in Figure 2.2 for the sake of clarity. Its order of magnitude is largely higher than the contribution of the other subgrid terms. It would require more specific attention in future works.

2.4.3. *Comparison of turbulence models*

After identifying the predominant terms, a comparison of the different turbulence models is proposed. Five models are compared: Smagorinsky's model (Smagorinsky 1963), the wall-adapting local eddy-viscosity (WALE) model (Nicoud and Ducros 1999), Bardina's model (Bardina *et al.* 1980), the Approximate Deconvolution Model (ADM) (Adams and Stolz 2002) and the mixed Smagorinsky–Bardina's model (Bardina *et al.* 1983). ADM is taken with an order of 6. This choice will be discussed here after.

For this comparison, the relative error of each model is evaluated using the L2 norm:

$$\epsilon_{conv}^{ADM} = \sqrt{\frac{\sum_{I,\ \varepsilon_k^* > 1\cdot 10^{-10}} \left(\bar{\tau}_{conv}^{DNS} - \bar{\tau}_{conv}^{ADM}\right)^2}{\sum_{I,\ \varepsilon_k^* > 1\cdot 10^{-10}} \left(\bar{\tau}_{conv}^{DNS}\right)^2}} \quad [2.14]$$

Figure 2.3 presents these relative errors for each model and each subgrid term with the most refined mesh. Since Smagorinsky's and the WALE models are mainly based on the consideration of kinetic energy and inertial effects in turbulent flows, they can only be applied to τ_{conv} and τ_{interf}. This figure highlights that ADM is the most appropriate model for all of the subgrid terms, whatever the phase (Vincent *et al.* 2016). However, for the water phase, the error level of ADM applied to the pressure subgrid term remains high. This has a limited effect since the pressure term is not predominant, as shown in the previous section.

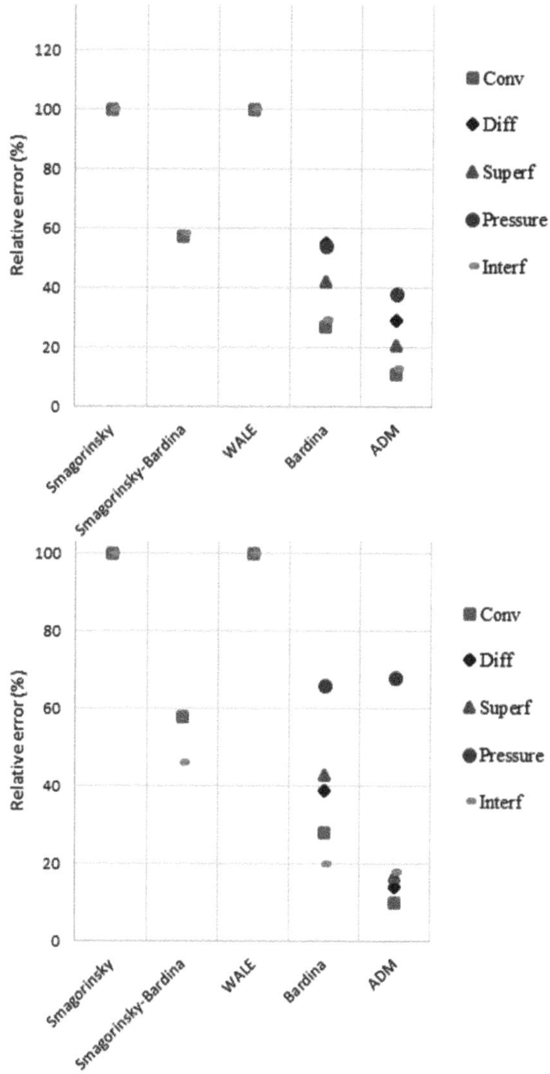

Figure 2.3. *Relative error obtained by comparison between the modeled subgrid term and the terms obtained by DNS, top: for the oil phase, bottom: for the water phase, only the z component is displayed, mesh with* 512^3 *cells. For a color version of this figure, see www.iste.co.uk/prudhomme/fluid1.zip*

Figure 2.4 (left) displays the dispersion and the slope obtained between each model and the filtered DNS results for the convective subgrid term. ADM is the only model to present a slope close to 1 with a limited dispersion.

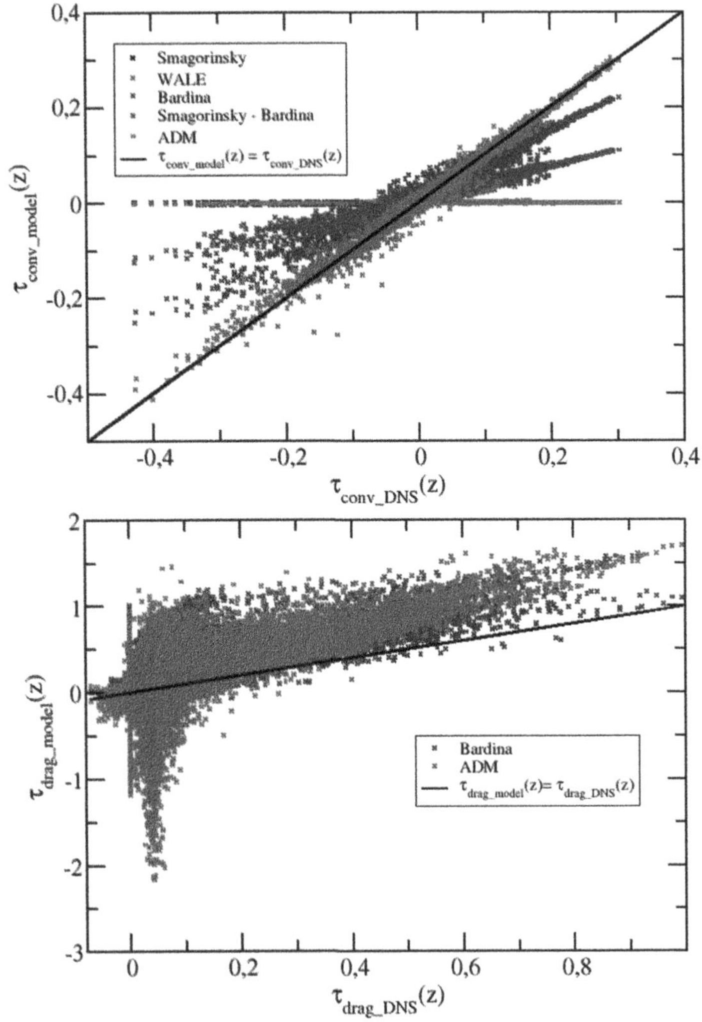

Figure 2.4. *Correlation between the turbulence models and the subgrid terms evaluated by DNS, top: convective subgrid term, bottom: drag subgrid term in the intermediate region $0.3 \leq \varepsilon_2^* \leq 0.7$, only the component in the z-direction is considered, mesh with 128^3 cells, oil phase. For a color version of this figure, see www.iste.co.uk/prudhomme/fluid1.zip*

No data is given in Figure 2.3 for the drag subgrid term, since the error is always higher than 100%. As shown in Figure 2.4 (right), this is probably due to the region splitting of the drag force expression, which induces a deviation of the modeled subgrid term at the boundaries (e.g. Sagaut and Germano 2005; Carrara and DesJardin 2008). A solution to model this subgrid term could be to adapt the phase-conditioned filtering proposed by Trontin *et al.* (2012) for the velocity and stress tensor jump at interfaces to the regions defined in the drag force expression.

To assess this assumption, a derivative expression of the drag force without region splitting is tested:

$$\mathbf{F}_{\mathbf{mix}} = \frac{1}{2}\overset{*}{\varepsilon}_1\overset{*}{\varepsilon}_2\left(\frac{18\mu_1}{\overset{*}{\varepsilon}_1 d^2_{pbubble}} + \frac{18\mu_2}{\overset{*}{\varepsilon}_2 d^2_{pdroplet}}\right)(\boldsymbol{u}_1 - \boldsymbol{u}_2) \qquad [2.15]$$

with $d_{pbubble} = \sqrt{\frac{\mu_2}{\mu_1}\frac{\overset{*}{\varepsilon}_2}{\|\nabla\overset{*}{\varepsilon}_2\|}}$ and $d_{pdroplet} = \sqrt{\frac{\mu_2}{\mu_1}\frac{\overset{*}{\varepsilon}_1}{\|\nabla\overset{*}{\varepsilon}_1\|}}$.

This drag force is not suitable for the simulation of three fields, but can be a first approximation in case of two continuous field simulations, such as in the phase inversion benchmark. With this derivative expression, the error for ADM decreases from more than 100%, with the splitting version to 69% with the non-splitting one and the most refined mesh.

To finish with the model comparison, the ability of each turbulence model to reproduce the effective viscosity predicted by DNS is studied. For this purpose, the equivalent viscosity, whose expression is given in equation [2.16], is compared with the turbulent viscosity of each model defined in equation [2.17].

$$\mu_{eq} = \overset{*}{\varepsilon}_1\frac{\bar{\tau}^{DNS}_{conv1} : \nabla\overline{\boldsymbol{u}_1}}{\overline{S_1} : \nabla\overline{\boldsymbol{u}_1}} + \overset{*}{\varepsilon}_2\frac{\bar{\tau}^{DNS}_{conv2} : \nabla\overline{\boldsymbol{u}_2}}{\overline{S_2} : \nabla\overline{\boldsymbol{u}_2}} \qquad [2.16]$$

$$\mu_{ADM} = \overset{*}{\varepsilon}_1\frac{\bar{\tau}^{ADM}_{conv1} : \nabla\overline{\boldsymbol{u}_1}}{\overline{S_1} : \nabla\overline{\boldsymbol{u}_1}} + \overset{*}{\varepsilon}_2\frac{\bar{\tau}^{ADM}_{conv2} : \nabla\overline{\boldsymbol{u}_2}}{\overline{S_2} : \nabla\overline{\boldsymbol{u}_2}} \qquad [2.17]$$

The results for Smagorinsky's model and ADM are displayed in Figure 2.5 for the most refined mesh (512^3 cells). The equivalent viscosity can be negative, especially close to interfaces. This observation was previously made by Labourasse *et al.* (2007). Smagorinsky's model is unable to reproduce these negative contributions and the other variations observed along the box height. On the contrary, a good agreement is obtained with ADM.

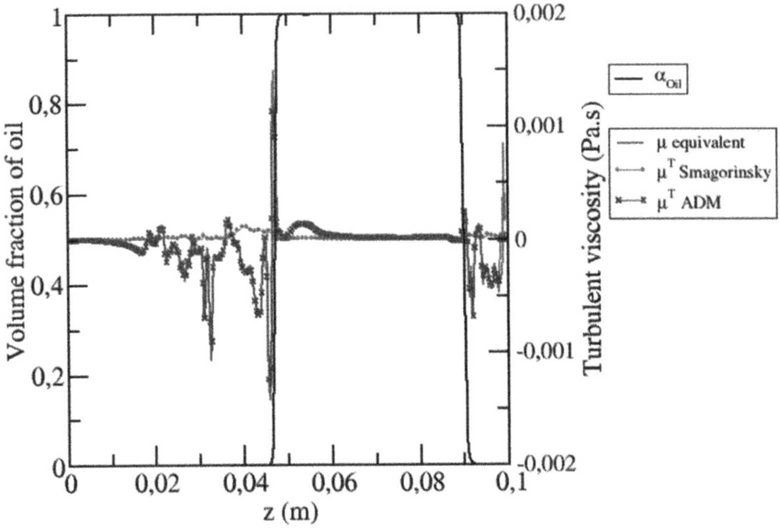

Figure 2.5. *Comparison of the equivalent viscosity μ_{eq} predicted by DNS and the turbulent viscosity obtained with Smagorinsky's model and ADM, the volume fraction of oil is also represented to locate interfaces, the viscosities are extracted from a slice of the domain at x = y = 8 cm, mesh with 512^3 cells. For a color version of this figure, see www.iste.co.uk/prudhomme/fluid1.zip*

2.4.4. *ADM order*

In all of these comparisons, ADM with an order of 6 has been used, since it corresponds to the best results obtained with the *a priori* LES study of the test case using a single-fluid approach (e.g. Tavares *et al.* 2016). However, in this section, a comparison of the relative error for the convection term and the deconvolution error on the velocity field is proposed according to the ADM order. The error for the convection term is evaluated using equation [2.14] and the deconvolution error with:

$$\epsilon_{u_k}^{ADM} = \sqrt{\frac{\sum_{\overline{\epsilon_k^*}^I > 1\cdot 10^{-10}} \left(\overline{u}_k^{DNS} - \overline{u}_k^{ADM}\right)^2}{\sum_{\overline{\epsilon_k^*}^I > 1\cdot 10^{-10}} \left(\overline{u}_k^{DNS}\right)^2}} \qquad [2.18]$$

The results are given in Figure 2.6 for the three grids. Contrary to the single-fluid formulation, if the order is increased, the accuracy is not automatically increased. There is an optimal value. Under this value, the increase of the order improves the deconvolution of the velocity field and so the relative errors decrease.

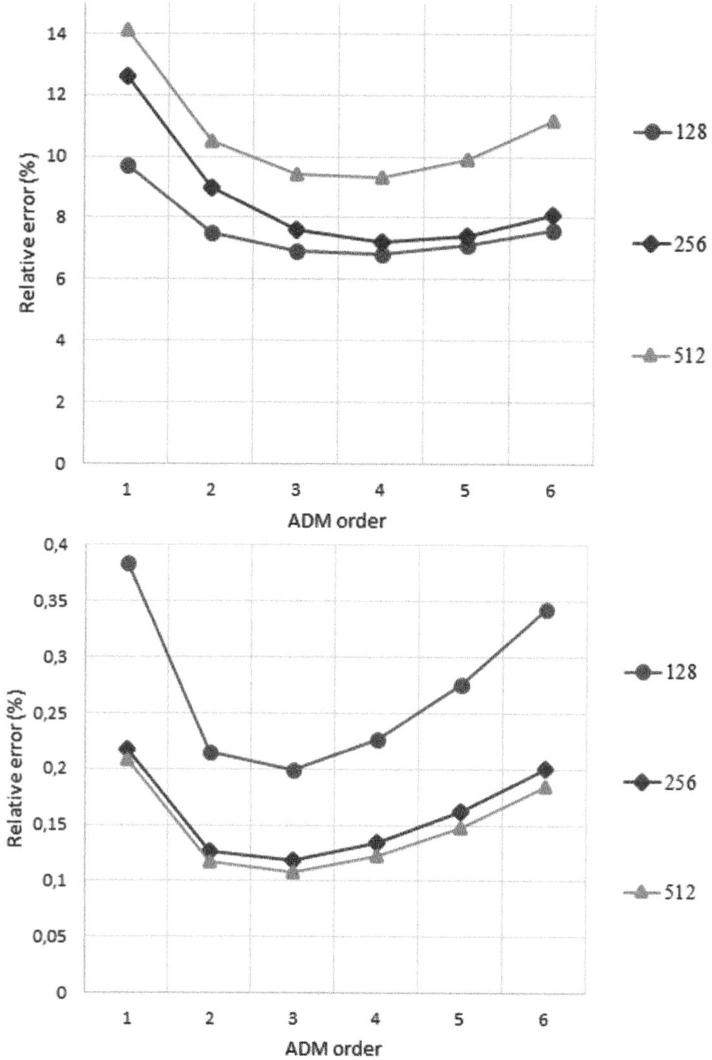

Figure 2.6. *Relative error of the convection subgrid term (top) and the predicted velocity field (bottom), according to the ADM order for the three grids and the oil phase. For a color version of this figure, see www.iste.co.uk/prudhomme/fluid1.zip*

However, when the order is increased after this optimal value, the ADM approximation of the velocity includes values of the considered field in cells that belong to a far neighborhood. The flow in this far neighborhood can be quite different from the cell that is of interest. This decorrelation has a negative effect and thus increases the deconvolution error. The optimal order is obtained when an equilibrium is reached between the positive effect of the order increase on the deconvolution approximation, and the negative effect of the decorrelation of the cells in the far neighborhood. For the convection term, this optimal value is equal to 4. We can note that the conclusions are the same if we are considering another subgrid term. Concerning the deconvolution error, the best order is equal to 3 for the three grids. Finally, we see that the deconvolution error is very small compared to the ADM error of the subgrid terms (see Figure 2.3). This error decreases with the mesh refinement, which means that the more refined the mesh is, the better the velocity field is approximated by ADM, whatever its order.

2.4.5. *Effect of the filter*

In this section, an analysis of the filter choice is proposed. The previous box filter (equation [2.19]) is compared to a Gaussian filter (equation [2.20]).

$$G(x) = \frac{1}{\widehat{\Delta}} h\left(\frac{\widehat{\Delta}}{2} - |x|\right) \qquad [2.19]$$

where h is the Heaviside function and $\widehat{\Delta}$ is the filtered width equal to $2\Delta x$ for the two filters.

$$G(x) = \sqrt{\frac{6}{\pi \widehat{\Delta}^2}}\, exp\left(\frac{-6x^2}{\widehat{\Delta}^2}\right) \qquad [2.20]$$

Figure 2.7 presents a comparison of the subgrid term weight according to the filter. Table 2.4 orders the term according to its contribution with the Gaussian filter, and has to be compared with Table 2.3. The weight of the subgrid terms τ_{superf}, τ_{conv} and τ_{diff} decreases with the Gaussian filter, but increases for τ_{interf} and $\tau_{pressure}$. Globally, the subgrid term contribution is then smaller. This is an important result because if a subgrid term has a limited contribution in the balance equations, the errors made by modeling it will have a limited impact. Thus, it can be preferable to use a filter that limits the contribution of the subgrid terms.

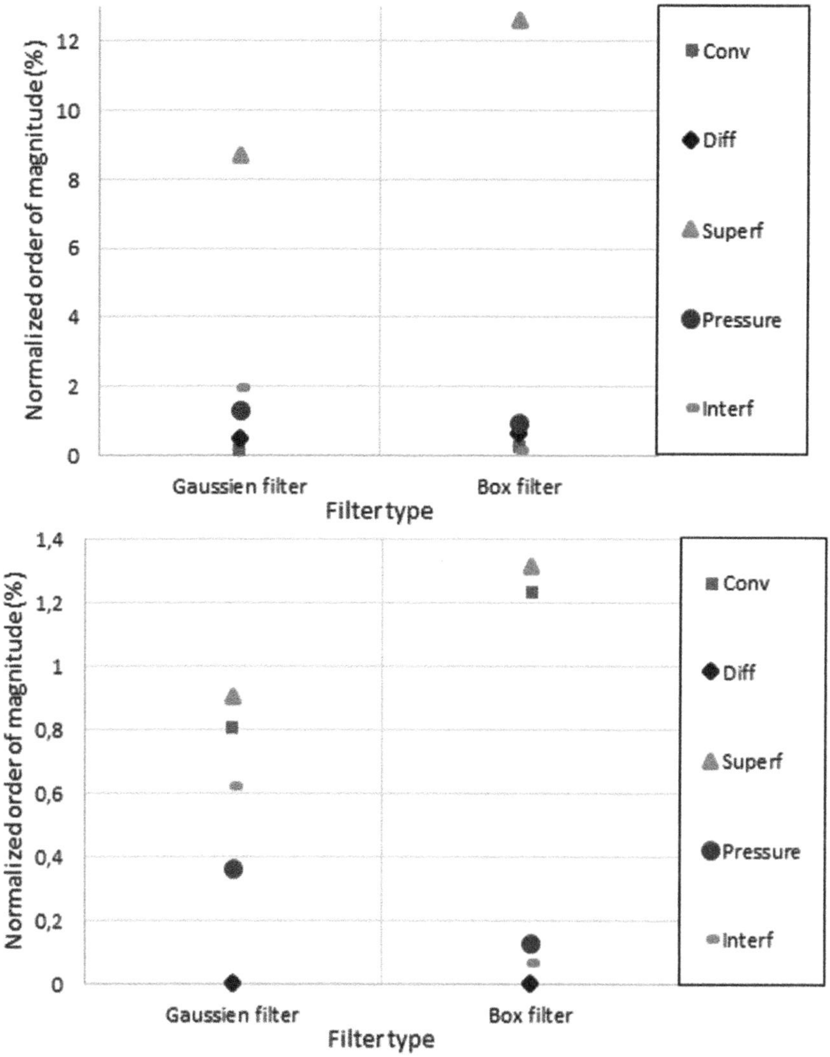

Figure 2.7. *Order of magnitude of the normalized subgrid terms obtained with the box and the Gaussian filter, top: for the oil phase, bottom: for the water phase, only the z component is displayed, mesh with 512^3 cells. For a color version of this figure, see www.iste.co.uk/prudhomme/fluid1.zip*

Then, the turbulence models are compared with the two filters in Figure 2.8 for the oil phase. The same results are obtained with the water phase. The filter has no effect on Smagorinsky's and the WALE models. These results were expected as these models are not strongly linked to the filter choice, contrary to the Bardina model and ADM. For these two models, the relative error decreases when the Gaussian filter is applied, except for the pressure subgrid term. Once again, the results are in agreement with the expectations. Indeed, in practice, applying a filter corresponds to evaluating a quantity in terms of its value over a few cells. With a box filter with a filter size of 2, the filtered quantity is evaluated by averaging all of the values over the cells, with at least one common node with the cell of interest. The surrounding cells are considered with the same weight. However, with the Gaussian filter, the same cells are considered, but they are weighted according to the distance from the cell of interest. This approach seems to be better, as the flow in the cells far from the cell of interest can be quite decorrelated from the flow in the cell of interest. This difference can be large enough to induce deconvolution errors and modeling inaccuracies.

	LES filter	
	Oil	Water
1	τ_{superf}	τ_{superf}
2	τ_{interf}	τ_{conv}
3	$\tau_{pressure}$	τ_{interf}
4	τ_{diff}	$\tau_{pressure}$
5	τ_{conv}	τ_{diff}

Table 2.4. *Classification of the subgrid terms according to their relative contribution for the 512^3 grid with the Gaussian filter*

The Gaussian filter then seems better than the box filter, since the contribution of the subgrid terms is globally smaller and the modeling errors are reduced.

According to the results presented in this section, in the following, ADM is implemented with an order of 3 using a Gaussian filter.

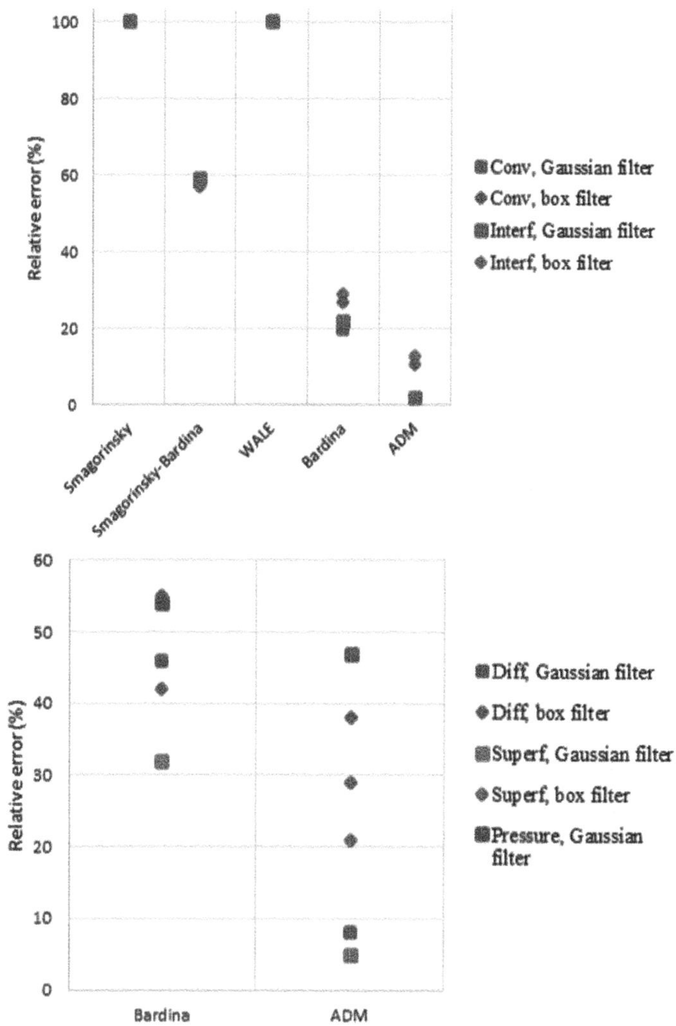

Figure 2.8. *Relative error obtained by comparison between the modeled subgrid term and the terms obtained by DNS with the box and the Gaussian filters, for the oil phase with the 512^3 grid. Only the z component is displayed. For a color version of this figure, see www.iste.co.uk/prudhomme/fluid1.zip*

2.5. Comparison of turbulence models with true LES

In the previous section, LES studies have been performed using filtered DNS. Here, true LES are presented on a two-phase flow using the multifield approach.

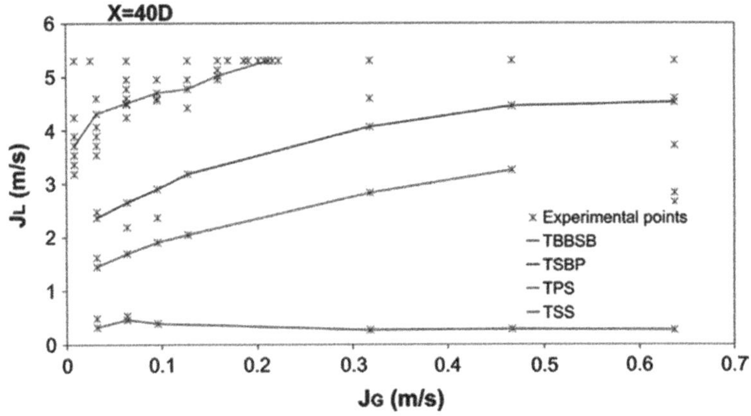

Figure 2.9. *METERO flow pattern at X = 40 D. J_L corresponds to the water velocity and J_G to the air velocity at injection. TSS refers to the transition from slug to stratified flow (pink line), TPS refers to the transition from plug to slug flow (orange line), TBBSB refers to the transition from buoyant bubble flow to stratified bubble flow (green line) and TSBP refers to the transition from the stratified bubble regime to plug (purple line). The crosses correspond to video acquisitions (e.g. Bottin* et al. *2014). For a color version of this figure, see www.iste.co.uk/prudhomme/fluid1.zip*

2.5.1. *Presentation of the METERO experiment*

The METERO experiment has been jointly developed in the framework of the NEPTUNE project by CEA, EDF, AREVA and IRSN, in order to improve numerical CMFD tools by contributing to model validations. This experiment allows us to establish the flow pattern map in Figure 2.9, according to the injection mass flow rate of water and air. The objective of the study is to show that LES modeling coupled with the multifield approach makes it possible to predict, qualitatively and quantitatively, the readjusted flow regimes. An ADM simulation is also performed first on this test case as a preliminary study.

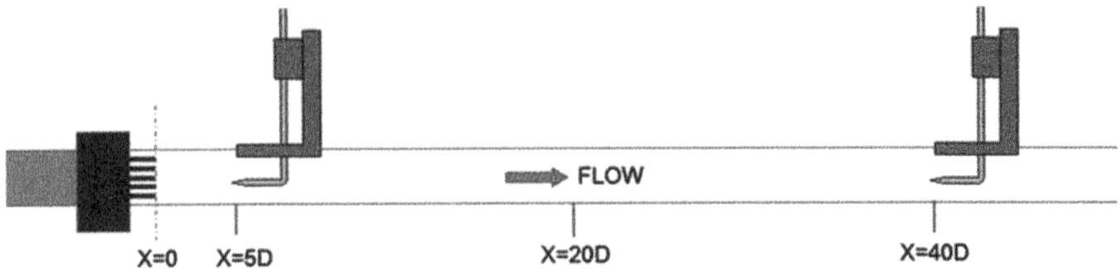

Figure 2.10. *Schematic view of the horizontal pipe of the METERO experiment (e.g. Bottin* et al. *2014)*

2.5.2. *Presentation of the test case*

The experiment consists of a horizontal cylindrical pipe (see Figure 2.10 for the schematic view), in which water and air can be injected at the same time with different mass flow rates (from 0 $m.s^{-1}$ to 5 $m.s^{-1}$ for water and from 0 $m.s^{-1}$ to 0.7 $m.s^{-1}$ for air). The air injection tubes have been set to ensure uniform bubble injection in the inlet section. The test section is 5.4 m long for an inner diameter of 0.1 m. The flow properties are extracted at three different positions along the pipe. Further information on the experiment can be found in, for example, Bottin *et al.* (2014).

2.5.3. *Simulation results*

The simulations of the experiment have been performed using a cylindrical mesh that contains 966,168 cells (see Figure 2.11). The boundary conditions are standard: inlet on the right, outlet on the left and walls everywhere else. The properties of the two phases are presented in Table 2.5. The surface tension coefficient is equal to 0.075 $N.m^{-1}$. The simulations are performed until they reach 13 s, including 5 s of averaging.

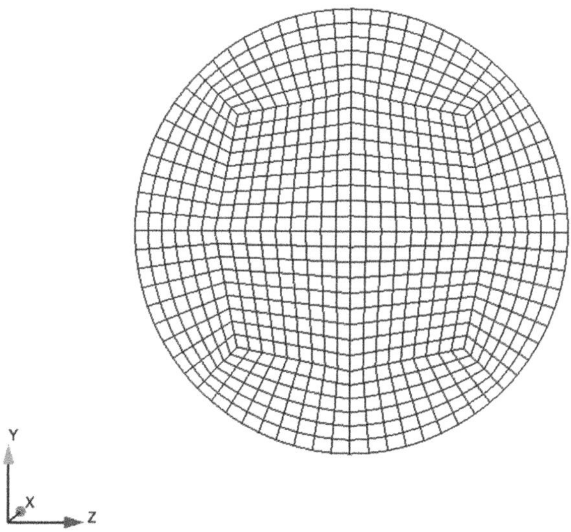

Figure 2.11. *Slice of the mesh used for the simulation of the METERO test case. For a color version of this figure, see www.iste.co.uk/prudhomme/fluid1.zip*

Phase	ρ (kg.m^3)	μ (kg.m^{-1}.s^{-1})
Air	1.29	$1.8 \cdot 10^{-5}$
Water	1000	$1.3 \cdot 10^{-3}$

Table 2.5. *Physical properties of the two phases in the METERO test case*

Flow regime	RANS	WALE
Dispersed bubble flow	0.5 ms	0.5 ms
Plug flow	0.25 ms	0.2 ms
Slug flow	0.12 ms	0.12 ms

Table 2.6. *Time steps according to the flow regime and the turbulence model for the METERO test case*

2.5.4. *Comparison between RANS and LES*

In the first study, the air velocity at the injection is fixed at 0.127 m.s^{-1}. Three liquid velocities are simulated: 4.42 m.s^{-1}, 2.65 m.s^{-1} and 1.59 m.s^{-1} corresponding, respectively, (according to Figure 2.9) to a stratified bubble flow, plug flow and slug flow. The time steps are given in Table 2.6. For these three simulations, the multifield approach is used. Thus, the small spherical bubbles are treated using the dispersed model. LBMo is activated for the large deformable bubbles. Three fields are defined: a continuous liquid field, a continuous gas field and a dispersed gas field. Isothermal transfer (break up and coalescence) between the two gaseous fields is allowed (e.g. Mimouni *et al.* 2016). The turbulence models are only applied to the liquid field. Figure 2.12 presents the qualitative comparison of the simulation results performed using the WALE model. When the liquid velocity decreases, the bubbles are supposed to become larger. Thus, the number of bubbles treated as a dispersed field decreases.

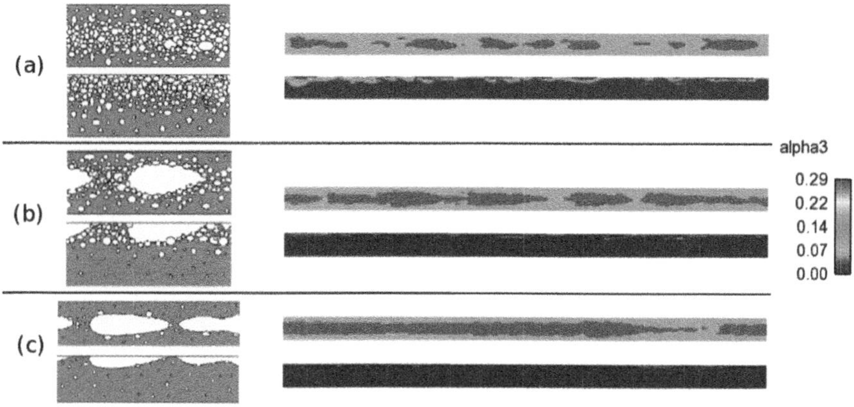

Figure 2.12. *Qualitative comparison of the simulations performed with the WALE model, (a) stratified bubble flow (J_L = 4.42 m.s^{-1}), (b) plug flow (J_L = 2.65 m.s^{-1}), (c) slug flow (J_L = 1.59 m.s^{-1}) and J_G = 0.127 m.s^{-1}. Left column: schematic view of the expected flow regime (top view and side view), right column: simulation results, top: top view of the isosurface of the continuous gas field, bottom: volume fraction of the dispersed gas field. For a color version of this figure, see www.iste.co.uk/prudhomme/fluid1.zip*

The simulation results correlate with these experimental observations. Then, in Figure 2.13, the WALE model is quantitatively compared to the RANS approach. The prediction of the liquid velocity and the void fraction is improved by using an LES model. This result was also expected because the RANS approach is based on an ensemble average description of turbulence effects. In flows that only contain small spherical bubbles treated with a dispersed approach, this view is suitable. However, when large interfaces are resolved, the deterministic description, intrinsically brought about by LES modeling, is more appropriate.

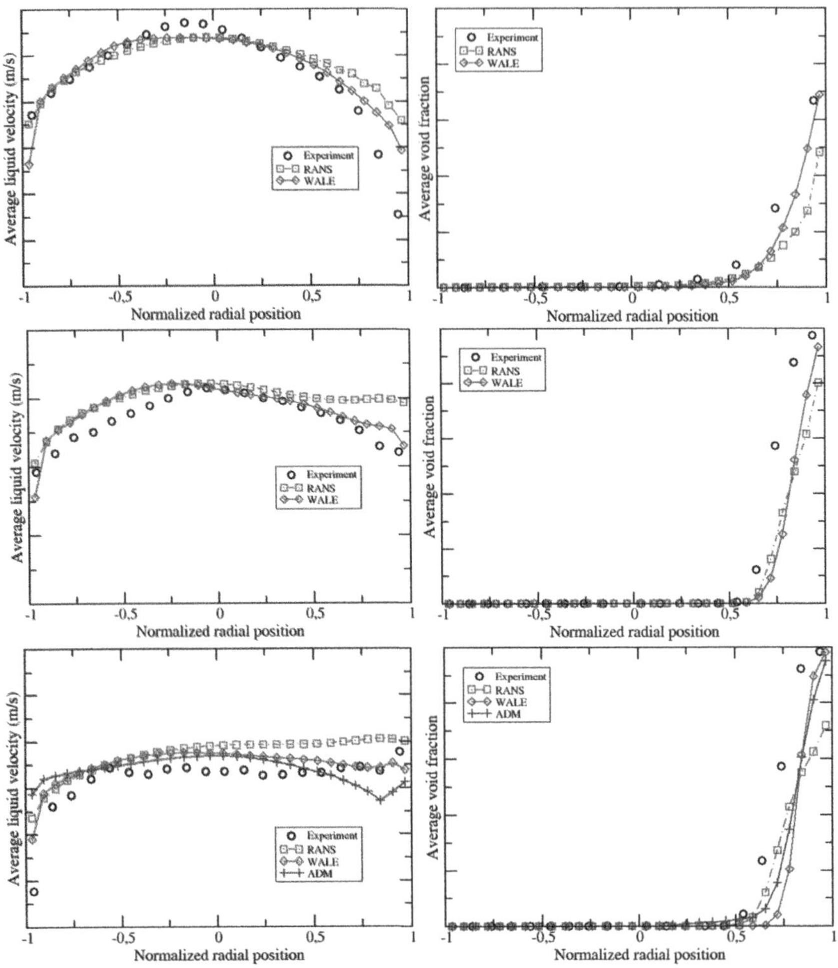

Figure 2.13. *Average liquid velocity and average void fraction at X = 40D. Top to bottom: stratified bubble flow (J_L = 4.42 m.s^{-1}), plug flow (J_L = 2.65 m.s^{-1}), slug flow (J_L = 1.59 m.s^{-1}) and J_G = 0.127 m.s^{-1}. For a color version of this figure, see www.iste.co.uk/prudhomme/fluid1.zip*

2.5.5. *ADM implementation*

To finish, a last simulation is performed using ADM. For this purpose, the slug flow test case (J_L = 1.59 m.s^{-1} and J_G = 0.127 m.s^{-1}) is simulated considering only two continuous fields. For this simulation, the ADM order is equal to 3 and reduced close to the walls. A Gaussian filter is used. To model the relaxation term (Schlatter *et al.* 2004), the dynamic Smagorinsky's model is applied (Germano *et al.* 1991). Moreover, a new definition is given for the filtered velocity. Indeed, in the two-fluid model, the velocity of a field k is defined in the whole domain, even if the field k is not present in one cell. In such cells, the value of the velocity of field k has no meaning and can be criticized. Thus, the following expression of the velocity is used for the filtering process:

$$\overline{\boldsymbol{u}_k}^{NEW} = \frac{\overline{\varepsilon_k^*}\,\overline{\boldsymbol{u}_k}}{max(\overline{\varepsilon_k^*}, 1\cdot 10^{-4})} \qquad [2.21]$$

With this expression, the smaller the volume fraction is, the smaller the velocity weight is in the filtering process.

To reconstruct the volume fractions with ADM, a treatment is made to correct the obtained values, such that the reconstructed volume fraction is between 0 and 1, which is not intrinsically ensured by the ADM process. Indeed, the reconstructed volume fraction has the following expression:

$$\overline{\varepsilon_k^*}^{ADM} = \sum_{l=0}^{N} \left(I_d - G\right)^l * \overline{\varepsilon_k^*} \qquad [2.22]$$

where N is the ADM order, I_d is the identity matrix, G is the LES filter, and $*$ is the convolution operator. With this expression, a simple analysis based on the barycenter theory shows that $\overline{\varepsilon_k^*} \in [0;1]$ does not imply that $\overline{\varepsilon_k^*}^{ADM} \in [0;1]$. However, if $\sum_k \overline{\varepsilon_k^*} = 1$, then $\sum_k \overline{\varepsilon_k^*}^{ADM} = 1$.

For this purpose, the negative value of $\overline{\varepsilon_k^*}^{ADM}$ is multiplied by -1. If $\overline{\varepsilon_{k1}^*}^{ADM}$ is negative, its value is multiplied by -1 and $\overline{\varepsilon_2^*}^{ADM}$ is readjusted, such that $\overline{\varepsilon_1^*}^{ADM} + \overline{\varepsilon_2^*}^{ADM} = 1$. This point is a major uncertainty in our implementation of ADM that would require further consideration. Finally, in a first attempt, only τ_{conv}, τ_{superf} and τ_{diff} are modeled.

The results, displayed in Figure 2.13, are encouraging. Indeed, they are not so far from the experimental data. Errors are probably made by only considering two continuous fields. It has to be noted that no discrepancies have been observed concerning mass conservation, since only $\overline{\varepsilon_k^*}^{ADM}$ was modified. The mass balance equation is solved considering $\overline{\varepsilon_k^*}$, which does not receive any special treatment. Moreover, in this first implementation, the subgrid term τ_{interf}, which appears in the mass balance equation, has not been modeled.

In future works, all of the choices made for the simulation should be assessed. Moreover, the extension to three field simulations should be carried out.

2.6. New phase change model for large interfaces

In the previous sections, the objective has been the investigation of LES to simulate the turbulence effects. Turbulence is a main issue for the simulation of complex flows evolving in nuclear power plants or in conventional heat exchangers. However, phase change also has a major impact. Within the multifield approach implemented in the framework of the two-fluid model, large interfaces are treated with the LBMo. Nevertheless, the approach does not include a model in the case of phase change at the interface. Thus, in this section, a model for the heat flux is proposed to deal with the phase change effects occurring at large interfaces.

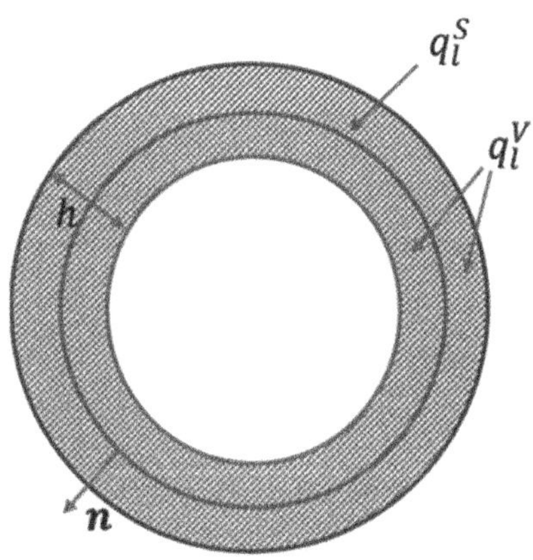

Figure 2.14. *Notations for the implementation of the new heat flux model. Red circle: interface with zero thickness, blue ring: interface with non-zero thickness. For a color version of this figure, see www.iste.co.uk/prudhomme/fluid1.zip*

2.6.1. *Implementation of the new heat flux model*

To obtain the new heat flux model, the same methodology proposed by Brackbill *et al.* (1992) for the surface tension force is used. For this purpose, we consider an interface with zero thickness, represented in red in Figure 2.14. For such interfaces, the heat flux expression for the liquid phase is:

$$q_l^S = \lambda_l \nabla T_l.\boldsymbol{n} \qquad [2.23]$$

For the demonstration here, only the liquid term is considered. Nevertheless, the same process can be followed to obtain the volume reformulation of the vapor term. As in for the surface tension model, the volume expression of the heat fluxes q_l^V corresponding to an interface with a thickness equal to h, can be obtained by considering:

$$\lim_{h\to 0} \int_{V^{Int}} q_l^V(x)\,dx^3 = \int_{A^{Int}} q_l^S\left(x^{Int}\right)dA \qquad [2.24]$$

where V^{Int} is the volume of an interface with a thickness equal to h and A^{Int} is the area of a zero thickness interface. To obtain the expression of $q_l^V(x)$, the definition of $q_l^S(x^{Int})$ is used:

$$\int_{A^{Int}} q_l^S(x^{Int})dA = \int_{V^{Int}} q_l^S(x)\ \delta\,(\boldsymbol{n}(x^{Int}).(x\text{-}x^{Int})dx^3 \qquad [2.25]$$

$$= \int_{V^{Int}} \lambda_l \nabla T_l(x).\boldsymbol{n}(x)\delta(\boldsymbol{n}(x^{Int}).(x\text{-}x^{Int})dx^3$$

where δ is a Dirac function indicating interface.

In Brackbill *et al.* (1992), the interface is located with a color function C, using a VOF approach. Thus, by introducing the notations of this chapter, we can write the following relation between the color function gradient of the diffused interface and the normal vector $\boldsymbol{n}$:

$$\lim_{h\to 0} \nabla C(x) = \boldsymbol{n}(x)\delta(\boldsymbol{n}\left(x^{Int}\right).\left(x - x^{Int}\right)[C] \qquad [2.26]$$

with $[C]$ being the jump of the color function over the interface. Using this expression, we obtain:

$$\int_{A^{Int}} q_l^S(x^{Int})dA = \lim_{h\to 0} \int_{V^{Int}} \lambda_l \nabla T_l(x) \frac{\nabla C(x)}{[C]} dx^3 \qquad [2.27]$$

Therefore, the volume expression of the heat flux $q_l^V(x)$, with the color function C, is:

$$q_l^{\mathrm{V}}(x) = \lambda_l \nabla T_l(x) \frac{\nabla C(x)}{[C]} \quad [2.28]$$

In the two-fluid formulation, as previously done in the surface tension model, this expression is rewritten:

$$q_k^V(x) = \overset{*}{\varepsilon_k} \lambda_k \nabla T_k(x) \nabla \overset{*}{\varepsilon_k}(x) \quad [2.29]$$

where $\nabla T_k(x) = \frac{T_{sat}\text{-}T_k(x)}{d^{Int}}$ and d^{Int} are the distance to the interface.

2.6.2. *Validation of the new heat transfer model*

To validate the new heat transfer model, two test cases involving steam and water are simulated. In all of these simulations, the interface motion is only driven by heat transfer. Mesh sensitivity is also studied for each test case.

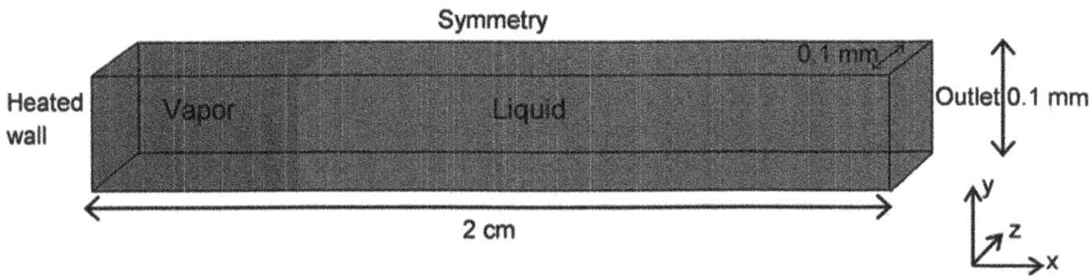

Figure 2.15. *Definition sketch of the 1D computational domain used for the simulation of the sucking and the Stefan problems. For a color version of this figure, see www.iste.co.uk/prudhomme/fluid1.zip*

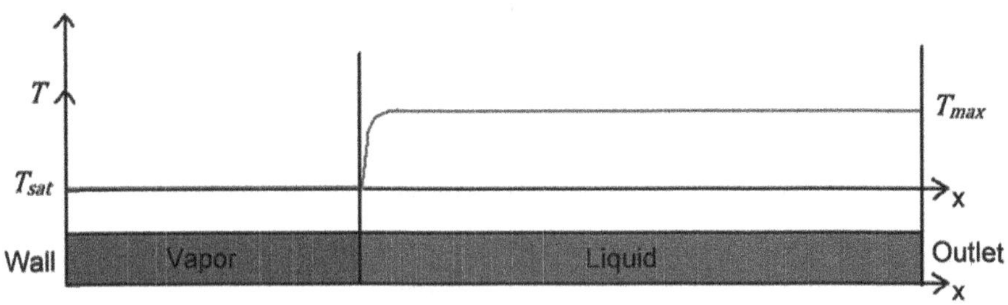

Figure 2.16. *Simulation conditions at a given time for the sucking problem, the liquid boils and the interface moves to the right side due to the volume expansion of the liquid–vapor phase change. For a color version of this figure, see www.iste.co.uk/prudhomme/fluid1.zip*

2.6.3. *Sucking problem*

In the sucking problem, steam and liquid water are contained in a 1D tube with a heated wall (see Figure 2.15). The liquid is superheated at T_{max} and the wall temperature is fixed at $T_{wall} = T_{sat}$, so that the steam is at the saturation temperature (see Figure 2.16). Therefore, there is no temperature gradient in the vapor. Both fields are at the rest at t = 0 s. For t > 0 s, the liquid begins to boil at the interface, which induces a displacement of the steam/water interface.

The theory of the sucking problem is given in Welch and Wilson (2000). First, let us make a transformation of the spatial coordinates, such as the steam/water is located at $\xi = 0$:

$$\xi = x - \int_0^t v_S(t')dt' \qquad [2.30]$$

where v_S is the interface velocity. With this new coordinate, the temperature profile is given by:

$$T = T_{sat} + B\sqrt{\frac{\pi}{2}}\phi'(0)\ exp(x_0^2)\ erf(x_0, x) \qquad [2.31]$$

where $B = \frac{\chi_l \rho_l}{C\rho_v}$, $C = \frac{\lambda_l}{L\rho_v}$, and χ_l are the thermal diffusivity of the liquid phase, L is the latent heat between steam and liquid water and $x = \frac{\eta + \phi'(0)}{\sqrt{2}}$, $x_0 = \frac{\phi'(0)}{\sqrt{2}}$, $\eta = \frac{\xi}{\sqrt{2\chi_l t}}$, and $erf(x_0, x) = \frac{2}{\sqrt{\pi}}\int_{x_0}^{x} \exp(-t^2)\,dt$ are the error function. The evolution of the interface position has the following expression:

$$X(t) = \frac{\phi'(0)\rho_l}{\rho_v}\sqrt{2\chi_l t} \qquad [2.32]$$

$\phi'(0)$ can be obtained by solving the following equation:

$$\phi'(0)exp\left(\frac{\phi'(0)^2}{2}\right) = \sqrt{\frac{2}{\pi}}\frac{T_{max} - T_{sat}}{B} \qquad [2.33]$$

The study is based on Sato's and Welch's publications (Welch and Wilson 2000; Sato and Niceno 2013). The physical properties of steam and water are updated at each time step using the standard set of thermodynamic

Equations Of State based on CATHARE functions (Emonot *et al.* 2011). The pressure of the system is equal to $1.013 \cdot 10^5$ Pa (the atmospheric pressure), and the liquid temperature in the bulk is $T_{max} = 378.15\ K$.

The tube length is equal to 2 cm. A heated wall boundary condition ($T_{wall} = T_{sat}$) is imposed on the left face, an outlet boundary condition is imposed on the right face and symmetry planes are imposed everywhere else (see Figure 2.15). The liquid temperature is initialized using equation [2.31] at t = 0.1 s. Therefore, the interface position at the beginning of the simulation is compared to the theoretical position at t = 0.1 s. Four different mesh refinements have been used with cell sizes equal to: $1 \cdot 10^{-4}$ m, $5 \cdot 10^{-5}$ m, $2.5 \cdot 10^{-5}$ m and $1.25 \cdot 10^{-5}$ m. The time step is kept constant and fixed according to the grid, to ensure a constant Fourier number in all of the simulations. The time steps are, respectively, equal to $1.6 \cdot 10^{-3}$ s, $4 \cdot 10^{-4}$ s, $1 \cdot 10^{-4}$ s and $2.5 \cdot 10^{-5}$ s.

Initially, steam is defined in the first four cells with a concentration of steam equal to 1. The next five cells are composed of a mixture of the two phases (concentration between 0 and 1) and the rest of the domain is filled with liquid (concentration equal to 0). The five mixed cells have the following composition in steam: 87.5% for the first one, 75%, 50%, 25% and 12.5%. This initialization is consistent with the five cells' interface thickness imposed by the interface sharpening equation.

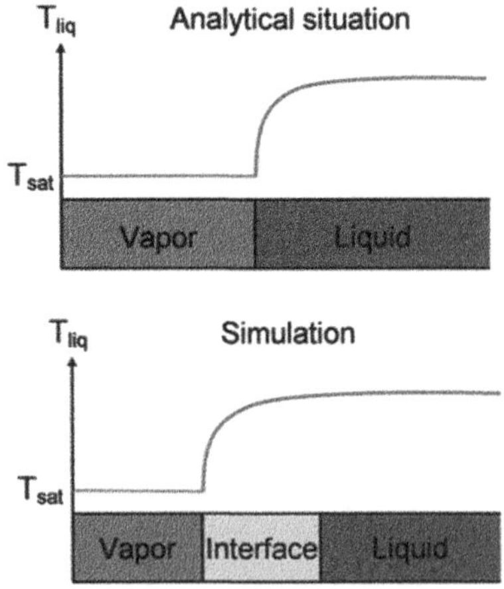

Figure 2.17. *Schematic view of the temperature profile and vapor/liquid interface in the analytical test case (top) and the simulation (bottom), sucking problem. For a color version of this figure, see www.iste.co.uk/prudhomme/fluid1.zip*

To compare the analytical solution with the simulation results, special care has to be taken concerning the interface and the temperature profile position. Indeed, in the analytical view of the test case (see Figure 2.17), the interface has a zero thickness and the temperature jump corresponds to the interface position. In the simulation, the temperature jump begins where the liquid is present, that is, at one extremity of the interface (see Figure 2.17). In the post-processing procedure, the position of the interface is taken where $\varepsilon_v^* = \varepsilon_l^* = 0.5$, corresponding to the center of the smeared interface. Since we consider that the analytical interface position corresponds to the position of the temperature jump, we have to shift the simulated results. Thus, an initial time onset is evaluated using the analytical expression of the interface position over time, and taking the analytical interface position at 0.1 s (the temperature profile is initialized at 0.1 s) decreased by $2.5\Delta x$ (five cells thickness). The simulated results are shifted using this time onset. Moreover, the interface position at 0.1s in all of the curves is considered as the zero position.

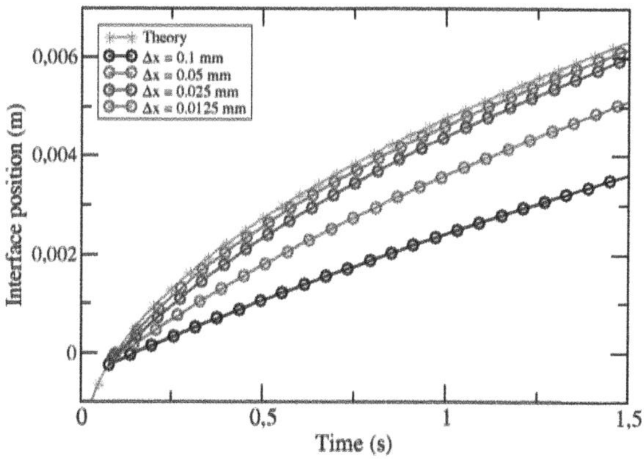

Figure 2.18. *Evolution of the interface position obtained with the new heat transfer model and different mesh refinements, sucking problem at an atmospheric pressure. For a color version of this figure, see www.iste.co.uk/prudhomme/fluid1.zip*

The results are given in Figure 2.18. The more refined the mesh is, the closer to the analytical solution the results are. The results obtained with the most refined mesh are very close to the analytical results. This confirms the ability of the developed model to predict an interface motion that is only driven by the heat transfer term. Figure 2.19 presents the corresponding average relative error for the interface position between 0.1 s and 1.5 s. When the mesh is refined, the error decreases. This study allows us to define the order of convergence ξ (using Richardson's extrapolation [2.31]) equal to 1.1.

$$\xi = \frac{ln\left(\frac{X_{m_3}-X_{m_2}}{X_{m_2}-X_{m_1}}\right)}{ln\left(\frac{1}{2}\right)} \quad [2.34]$$

with m_1, m_2 and m_3 as the three mesh refinements, such that:

$$\Delta x_{m_3} = \frac{\Delta x_{m_2}}{2} = \frac{\Delta x_{m_1}}{4} \quad [2.35]$$

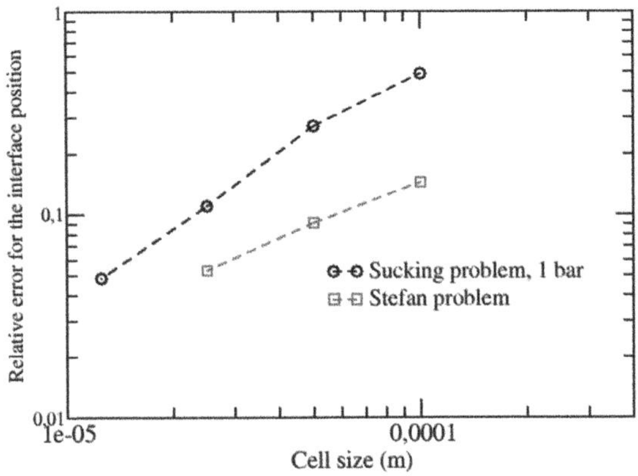

Figure 2.19. *Average relative error for the interface position compared to the theoretical results obtained with the new heat flux model for the two test cases presented in this chapter. For a color version of this figure, see www.iste.co.uk/prudhomme/fluid1.zip*

2.6.4. *Stefan problem*

The Stefan problem is very close to the sucking problem. Steam and liquid water are contained in a 1D tube with a heated wall under atmospheric pressure. However, in this case, steam is superheated, and the liquid water temperature is equal to the saturation temperature $T_{sat} = 373.15\ K$. The wall temperature keeps a constant value during the simulation: $T_{wall} = 398.15\ K$. As a result, there is a temperature gradient in the vapor phase (see Figure 2.20).

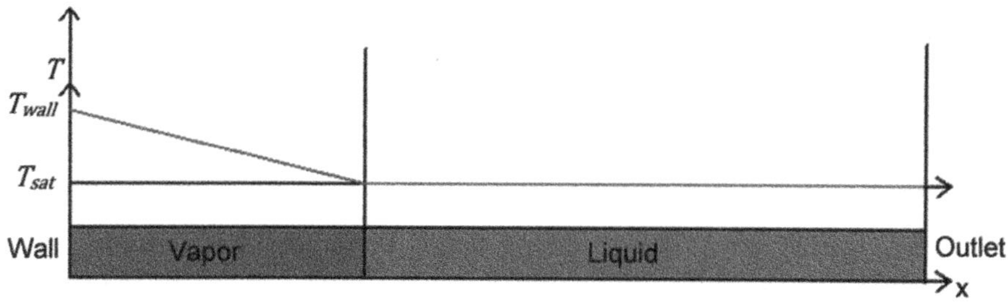

Figure 2.20. *Simulation conditions at a given time for the Stefan problem, the liquid boils and the interface moves to the right side due to volume expansion of the liquid-vapor phase change. For a color version of this figure, see www.iste.co.uk/prudhomme/fluid1.zip*

The evolution of the interface position is given by the equation (Welch and Wilson 2000):

$$X(t) = 2\beta\sqrt{\chi_v t} \qquad [2.36]$$

The temperature profile is equal to:

$$T(x,t) = T_{wall} + \frac{T_{sat}-T_{wall}}{erf(\beta)}\, erf\left(\frac{x}{2\sqrt{\chi_v t}}\right) \qquad [2.37]$$

where β is the solution of the equation:

$$\beta \exp(\beta^2)\, erf(\beta) = \frac{C_{pv}(T_{wall}-T_{sat})}{\sqrt{\pi}L} \qquad [2.38]$$

with C_{pv} as the specific heat capacity of vapor.

The physical properties of steam and water are updated at each time step using the standard set of thermodynamic Equations Of State based on CATHARE functions (e.g. Emonot *et al.* 2011). The pressure of the system is equal to $1.013 \cdot 10^5$ Pa. The same grids have been used with the same boundary conditions. The only difference is the wall temperature, which is equal to 398.15 K, that is, $T_{sat} + 25\,K$, in the Stefan problem, instead of T_{sat} in the sucking problem. Three different mesh refinements have been used with cell sizes equal to: $1 \cdot 10^{-4}$ m, $5 \cdot 10^{-5}$ m and $2.5 \cdot 10^{-5}$ m. The time steps are, respectively, equal to $1.6 \cdot 10^{-3}$ s, $4 \cdot 10^{-4}$ s and $1 \cdot 10^{-4}$ s.

The volume fractions are initialized, as done previously with the sucking problem. The initial vapor temperature is approximated using an affine function:

$$T_v(x) = T_{wall} + \frac{T_{sat}\text{-}T_{wall}}{d}\, x \qquad [2.39]$$

with $d = 9\Delta x$ (four cells full of vapor and five mixed cells), which corresponds to the "end" of the interface (see Figure 2.21), and x as the position of the cell center. Concerning the comparison of the simulation results and the analytical solution, the same observations can be done with the location of the interface and temperature deflection, as shown in Figure 2.21. The time onset is calculated using the initial computed interface center position increased by $2.5\Delta x$.

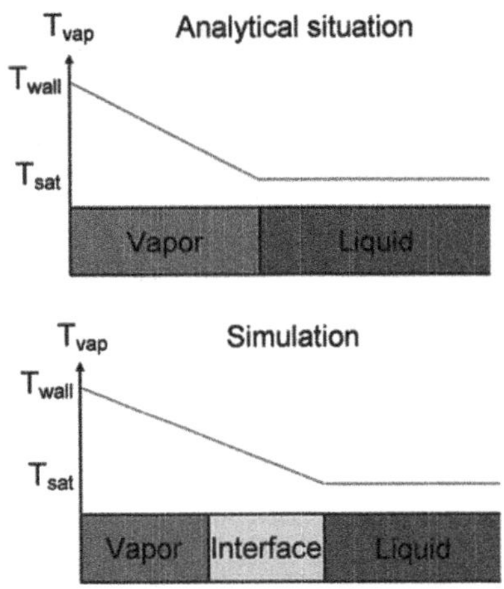

Figure 2.21. *Schematic view of the temperature profile and vapor–liquid interface in the analytical test case (top) and the simulation (bottom), Stefan problem. For a color version of this figure, see www.iste.co.uk/prudhomme/fluid1.zip*

The results are given in Figure 2.22. Contrary to the sucking problem, the improvement due to the mesh refinement is limited. This result can be explained by the fact that the local temperature gradient at the interface is smaller, even if in the whole domain, the temperature difference is higher (equal to 25 K versus 5 K in the sucking problem). Figure 2.19 presents the corresponding average relative error for the interface position for the first 10 s. When the mesh is refined, the error decreases. Thus, a mesh convergence is observed. The order of convergence (using Richardson's extrapolation [2.34]) is equal to 0.74.

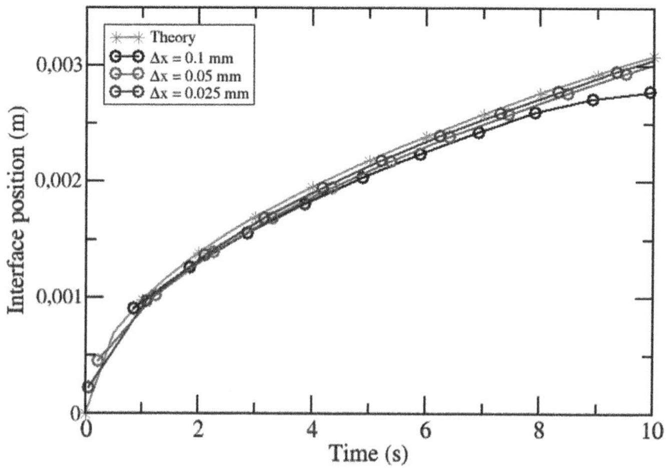

Figure 2.22. *Evolution of the interface position obtained with the new heat transfer model and different mesh refinements, Stefan problem at an atmospheric pressure. For a color version of this figure, see www.iste.co.uk/prudhomme/fluid1.zip*

The sucking and the Stefan problems are interesting test cases, as they can be used to calibrate the empirical models usually used to deal with phase changes in commercial codes at different pressure and temperature conditions for water/steam flows, as an example.

In this chapter, two 1D test cases have been used to validate the new heat transfer model considering super-saturated liquid water, on one hand, and super-saturated steam, on the other. The validation has been successfully extended to 3D simulations, under-saturated water and pressure conditions occurring in a reactor vessel or a steam generator of a nuclear power plant (e.g. Fleau 2017).

2.7. Conclusion

A multifield approach based on a two-fluid model has been implemented in the CMFD code NEPTUNE_CFD, to simulate complex flows containing inclusions with a large range of sizes. This approach has been widely validated in previous papers on laminar isothermal flows. However, turbulence effects with the two-fluid model applied to large and deformable interfaces have never been studied. Thus, the equations have been filtered for the first time. Two specific subgrid terms appeared conducting to a total of seven subgrid terms. The analysis of the weight of each term highlighted that the surface tension subgrid term was predominant, in agreement with previous LES studies on one-fluid models, whereas the diffusive subgrid term could not be negligible. The comparison of turbulence models finally showed that ADM was the most suitable model for all subgrid terms appearing in the filtered equations. Then, LES models were used to predict flow regimes. Better results were observed with LES, since the RANS approaches underestimated void fractions. The simulation performed with ADM was encouraging. Work has to be done to assess the different modeling choices, improve the results and clarify the sense of filtering the drag terms in the momentum equations. Finally, after investigating turbulence, phase change at large interfaces have been explored. For this purpose, a new heat flux model has been implemented and validated in this chapter.

The main perspective of this work concerns the implementation of ADM in two-phase flows, with the treatment of interfaces and the implementation of all of the subgrid terms. Extensions to three field simulations with phase change effects are expected.

2.8. References

Adams, N. and Stolz, S. (2002). A subgrid-scale deconvolution approach for shock capturing. *J. Comput. Phys.*, 178(2), 391–426.

Alizadehdakhel, A. *et al.* (2010). CFD modeling of flow and heat transfer in a thermosyphon. *Int. Commun. Heat Mass Transfer*, 37(3), 312–318.

Anderson, T.B. and Jackson, R. (1967). Fluid mechanical description of fluidized beds: Equations of motion. *Ind. Eng. Chem. Fundam.*, 6(4), 527–539.

Bardina, J. *et al.* (1980). Improved subgrid scale models based on large-eddy simulation. *AIAA Paper*, 80.

Bardina, J. *et al.* (1983). Improved turbulence models based on large eddy simulation of homogeneous, incompressible, turbulent flows. Tech. rep., Thermosciences Division, Department of Mechanical Engineering, Stanford University.

Bartosiewicz, Y. *et al.* (2008). A first assessment of the NEPTUNE_CFD code: Instabilities in a stratified flow comparison between the VOF method and a two-field approach. *Int. J. Heat and Fluid Flow*, 29, 460–478.

Bottin, M. *et al.* (2014). Experimental investigation of a developing two-phase bubbly flow in horizontal pipe. *Int. J. Multiphase Flow*, 60, 161–179.

Brackbill, J.U. *et al.* (1992). A continuum method for modeling surface tension. *J. Comput. Phys.*, 100, 335–354.

Capecelatro, J. *et al.* (2010). Eulerian-Lagrangian simulations of three-dimensional turbulent riser flows. *Proceedings of the 63rd Annual Meeting of the APS Division of Fluid Dynamics*, Long Beach, California.

Carrara, M.D. and DesJardin, P.E., (2008). A filtered mass density function approach for modeling separated two-phase flows for LES II: Simulation of a droplet laden temporally developing mixing layer. *Int. J. Multiphase Flow*, 34, 748–766.

Denèfle, R. *et al.* (2015). Multifield hybrid approach for two-phase flow modeling – Part 1: Adiabatic flows. *Comput. Fluids*, 113, 106–111.

Drew, D.A. and Lahey, R.T. (1979). Application of general constitutive principles to the derivation of multidimensional two-phase flow equations. *Int. J. Multiphase Flow*, 5(4), 243–264.

Duponcheel, M. *et al.* (2017). Experimental and numerical investigations of a two-phase wavy flow. *Nucl. Eng. Des.*, 321, 199–218.

Emonot, P. *et al.* (2011). CATHARE-3: A new system code for thermal-hydraulics in the context of the NEPTUNE project. *Nucl. Eng. Des.*, 241, 4476–4481.

Fang, C. *et al.* (2010). Volume of fluid simulation of boiling two-phase flow in a vapor-venting microchannel. *Front. Heat Mass Transfer*, 1, 1–11.

Fleau, S. (2017). Multifield approach and interface locating method for two-phase flows in nuclear power plant. PhD Thesis, University of Paris-Est Marne-la-Vallée, EDF R&D.

Fleau, S. *et al.* (2015). Validation of a multifield approach for the simulations of two-phase flows. *Comput. Therm. Sci.*, 7(5–6), 441–457.

Fleau, S. *et al.* (2016). Study of an interface sharpening method for large interface tracking within a two-fluid model. *Proceeding of the 9th International Conference on Multiphase Flow: ICMF 2016*, Florence.

Ganapathy, H. *et al.* (2013). Volume of fluid-based numerical modeling of condensation heat transfer and fluid flow characteristics in microchannels. *Int. J. Heat Mass Transfer*, 65, 62–72.

Germano, M. *et al.* (1991). A dynamic subgrid-scale eddy viscosity model. *Phys. Fluids*, 3, 1760–1765.

Guelfi, A. *et al.* (2007). Neptune – A new software platform for advanced nuclear thermal hydraulics. *Nucl. Sci. Eng.*, 156, 281–324.

Herrmann, M. (2010). A parallel Eulerian interface tracking/Lagrangian point particle multi-scale coupling procedure. *J. Comput. Phys.*, 229, 745–759.

Hirt, C.W. and Nichols, B.D. (1981). Volume of fluid (VOF) method for the dynamics of free boundaries. *J. Comput. Phys.*, 39, 201–225.

Ishii, M. (1975). *Thermo-Fluid Dynamic, Theory of Two-Phase Flow*. Eyrolles, Paris.

Kunkelmann, C. *et al.* (2012). The effect of three-phase contact line speed on local evaporative heat transfer: Experimental and numerical investigations. *Int. J. Heat Mass Transfer*, 55, 1896–1904.

Labourasse, E. *et al.* (2007). Towards large eddy simulation of isothermal two-phase flows: Governing equations and *a priori* tests. *Int. J. Multiphase Flow*, 33, 1–39.

Lakehal, D. (2004). DNS and LES of turbulent multifluid flows. *Proceedings of 3rd International Symposium on Two-Phase Flow Modelling and Experimentation*, Pisa.

Larocque, J. *et al.* (2010). Parametric study of LES subgrid terms in a turbulent phase separation flow. *Int. J. Heat Fluid Flow*, 31, 536–544.

Lebas, R. *et al.* (2009). Numerical simulation of primary break-up and atomization: DNS and modelling study. *Int. J. Multiphase Flow*, 35(3), 247–260.

Lee, W. (1980). *A Pressure Iteration Scheme for Two-phase Flow Modeling, Multiphase Transport Fundamentals, Reactor Safety, Applications*. Hemisphere Publishing, Washington.

Ling, Y. *et al.* (2015). Multiscale simulation of atomization with small droplets represented by a lagrangian point-particle model. *Int. J. Multiphase Flow*, 76, 122–143.

Mao, W. (2009). Numerical simulation of vapor-liquid phase change heat transfer and micromixing in microfluidic systems. M.Sc. Thesis, Guangzhou Institute of Energy Conversion, Chinese Academy of Sciences, China.

Mimouni, S. *et al.* (2011). Combined evaluation of second order turbulence model and polydispersion model for two-phase boiling flow and application to fuel assembly analysis. *Nucl. Eng. Des.*, 241(11), 4523–4536.

Mimouni, S. *et al.* (2016). Towards a more general CFD modelling for all flow regimes, ICMF 2016. *Proceeding of the 9th International Conference on Multiphase Flow*, Florence.

Mimouni, S. *et al.* (2017). CFD calculations of flow pattern maps and LES of multiphase flows. *Nucl. Eng. Des.*, 321, 118–131.

Nichita, A. and Thome, J. (2010). A level set method and a heat transfer model implementation into FLUENT for modeling of microscale two-phase flows. *Poceedings of "AVT-178 Specialists" Meeting on System Level Thermal Management for Enhanced Platform Efficiency*, No. EPFL-CONF-151629.

Nicoud, F. and Ducros, F. (1999). Subgrid-scale stress modelling based on the square of the velocity gradient tensor. *Flow, Turbul. Combust.*, 62(3), 183–200.

Olsson, E. and Kreiss, G. (2005). A conservative level set method for two phase flow. *J. Comput. Phys.*, 210, 225–246.

Patankar, S. and Spalding, D. (1972). A calculation procedure for heat, mass and momentum transfer in three-dimensional parabolic flows. *Int. J. Heat Mass Transf.*, 15(10), 1787–1806.

Sagaut, P. and Germano, M. (2005). On the filtering paradigm for LES of flows with discontinuities, *J. Turbul.*, 6(23).

Sato, Y. and Niceno, B. (2013). A sharp-interface phase change model for a mass-conservative interface tracking method. *J. Comput. Phys.*, 249, 127–161.

Schepper, S.D. *et al.* (2009). Modeling the evaporation of a hydrocarbon feedstock in the convection section of a steam cracker. *Comput. Chem. Eng.*, 33, 122–132.

Schlatter, P. *et al.* (2004). LES of transitional flows using the approximate deconvolution model. *Int. J. Heat Fluid Fl.*, 25(3), 549–558.

Smagorinsky, J. (1963). General circulation experiments with the primitive equations. *Monthly Weather Review*, 91, 99–165.

Štrubelj, L. (2009). Numerical simulations of stratified two-phase flows with two-fluid model and interface sharpening. M.Sc. Thesis, University of Ljublana, Slovenia.

Sun, D. *et al.* (2014). Modeling of the evaporation and condensation phase-change problems with FLUENT. *Numer. Heat Transfer*, Part B, 66, 326–342.

Sussman, M. *et al.* (1994). A level set approach for computing solutions to incompressible two-phase flow. *J. Comput. Phys.*, 114, 146–159.

Tavares, M. *et al.* (2016). *A priori* study for the modeling of LES subgrid scale terms. *Proceedings of the 4th Turbulence and Interactions Conference TI2015*, Cargèse.

Tomar, G. *et al.* (2010). Multiscale simulations of primary atomization. *Comput. Fluids*, 39(10), 1864–1874.

Trontin, P. *et al.* (2012). A phase-conditioned filtering of incompressible interfacial multiphase flow equations: *A priori* study for the modeling of LES subgrid scale terms. *Proceedings of Turbulence and Interactions*, La Saline-les-Bains, La Réunion.

Unverdi, S.O. and Tryggvason, G. (1992). A front-tracking method for viscous, incompressible, multi- fluid flows. *J. Comput. Phys.*, 10, 25–37.

Vallet, A. and Borghi, R. (1999). Modélisation eulérienne de l'atomisation d'un jet liquide. *C. R. Acad. Sci. Paris*, Série IIb, 327(10), 1015–1020.

Vincent, S. *et al.* (2008). Numerical simulation of phase separation and *a priori* two-phase LES filtering. *Comput. Fluids*, 37(7), 898–906.

Vincent, S. *et al.* (2018). *A priori* filtering and LES modeling of turbulent two-phase flows application to phase separation. *Comput. Fluids.*, 176, 245–259.

Welch, S. and Wilson, J. (2000). A volume of fluid based method for fluid flows with phase change. *J. Comput. Phys.*, 160, 662–682.

Wu, H. *et al.* (2007). Simulation of refrigerant flow boiling in serpentine tubes. *Int. J. Heat Mass Transfer*, 50, 1186–1195.

Yang, Z. *et al.* (2008). Numerical and experimental investigation of two-phase flow during boiling in a coiled tube. *Int. J. Heat Mass Transfer*, 51(5), 1003–1016.

Zuzio, D. *et al.* (2013). Numerical simulation of primary and secondary atomization. *C. R. Mécanique*, 241, 15–25.

3

An Original Approach to Extract Momentum and Heat Transfers from Particle-Resolved Simulations of Particulate Flows

A viscous penalty method (Vincent *et al.* 2014) is used to simulate the interaction between spherical particles and the surrounding carrier fluid using particle-resolved direct numerical simulations. An original method, based on Aslam extensions (Aslam 2004), has been developed and validated in order to extract from these simulations the hydrodynamic forces and heat transfers on immersed boundaries representing the particles. This method is an improvement to a previous work based on Lagrange extrapolations Chadil *et al.* (2018a, c). Comparisons between these two approaches are considered on various incompressible particulate flows such as the flow around an isolated particle at various Reynolds numbers and flows across packed spheres under face-centered cubic mono- and bi-disperse arrangements. These comparisons show a notable improvement, especially in complex configurations.

3.1. Introduction

Particle-laden flows are widely encountered in nature as well as in various industrial processes. Among the wide variety of applied problems, we can cite volcanic eruptions (Dartevelle 2007; Bagheri *et al.* 2015), oil refining, blast furnaces or chemical looping combustion (Lyngfelt *et al.* 2001; Adanez *et al.* 2012; Mattisson *et al.* 2018), fluid catalytic cracking reactors (Ali and Rohani 1997), gas-phase polymerization reactors (Gobin *et al.* 2003) and fluidized

Chapter written by Mohamed-Amine CHADIL, Stéphane VINCENT and Jean-Luc ESTIVALÈZES.

beds (Vincent *et al.* 2014; Ozel *et al.* 2017). Their modeling and simulation at the application scale are generally based on statistical approaches where macroscopic Eulerian–Eulerian or mesoscopic Eulerian–Lagrangian models (the average interphase transfer of momentum and energy between the fluid and the particles) are needed to be modeled.

Numerous well-known drag force laws are classically used to model the momentum interphase transfer at large scales. The most popular is the Schiller and Naumann (1933) correlation for a single sphere, whereas for fixed and fluidized beds of spheres, the correlations proposed by Ergun (1952) and Wen and Yu (1966) are often used. Heat transfers have also been modeled extensively for a single sphere using the Ranz and Marshall (1952) correlation, as well as for random arrangements of spheres with the Gunn (1978) law, to name the most notable onees.

Particle-resolved direct numerical simulation (PR-DNS) has been the favored tool of many researchers in order to model unclosed terms in macroscopic statistical equations. Many numerical approaches have been developed to perform PR-DNS of gas–solid flows. These can be classified into two types of approaches. The first relies on body-fitted mesh to impose boundary conditions at the particle surface. Many codes use this technique, for instance AVBP code (Schönfeld and Rudgyard 1999) used by Massol (2004) to carry out simulations of various configurations of spheres past by a uniform flow. The second type of approach is called the fictitious domain method. This method employs a fixed Cartesian grid in the whole domain, where each phase (solid or fluid) is located thanks to the so-called phase or color function C, which equals 1 in the solid phase and 0 in the fluid phase. Amongst the many fictitious domain methods for handling finite-size particles on fixed meshes, we can cite the Lattice Boltzmann approach (Ladd 1994a, b; Hill *et al.* 2001; Hoef *et al.* 2005; Beetstra *et al.* 2007), the immersed boundary method (IBM) (Uhlmann 2005, 2008; Wang *et al.* 2008; Deen *et al.* 2012; Zastawny *et al.* 2012; Tavassoli *et al.* 2013) and the approach used in this chapter, the viscous penalty method (Vincent *et al.* 2014; Chadil *et al.* 2019), to name the most used methods in the literature.

In the fictitious domain methods, the treatment of the fluid/solid interface using Navier–Stokes equations results in contaminated solution fields that do not correspond to unmodified Navier–Stokes equations. In the case of the immersed boundary method, Tenneti *et al.* (2011) proposed the so-called

PUReIBM method, where the immersed boundary (IB) forcing is restricted to the Eulerian grid points that lie in the solid phase, ensuring that the flow solution in the fluid phase is uncontaminated by the IB forcing. This allows them to easily compute drag forces (Tenneti *et al.* 2011) and heat transfers (Tenneti *et al.* 2013; Sun *et al.* 2015). In the case of the viscous penalty method (Vincent *et al.* 2014) that we consider, we proposed, in a previous work (Chadil *et al.* 2018c), a Lagrange extrapolation for the drag force computation in order to reach the uncontaminated fluid area far from the particle/fluid interface, that way ensuring accurate results for drag coefficients. This method was also used in the heat transfer computation (Chadil *et al.* 2018a) and extended to ellipsoidal particles (Chadil *et al.* 2018b).

In the last decade, many researchers have conducted numerical studies to characterize, understand and model the momentum and energy interphase exchanges in gas–solid flows. Some of them used the Lattice Boltzmann approach; for example, Hill *et al.* (2001) used it to study the hydrodynamic force for monodisperse face cubic-centered (FCC) arrays of spheres, whereas Hoef *et al.* (2005) and Beetstra *et al.* (2007) used it for random monodisperse and bidisperse arrays of spheres to cite a few. Another class of very popular methods for handling finite-size particles is the immersed boundary method (IBM) coupled with incompressible Navier–Stokes equations initially developed for particulate flows by Uhlmann (2005, 2008) and Wang *et al.* (2008). Deen *et al.* (2012) and Tavassoli *et al.* (2013) used this for three-dimensional PR-DNS with inflow and outflow boundary conditions to compute the gas–solid Nusselt number that is compared with Gunn's correlation (Gunn 1978). Note that Gunn (1978) proposed his correlation based on experimental results. Tenneti *et al.* (2011) used the so-called PUReIBM method with periodic boundary conditions for a random monodisperse array of spheres. A new correlation was proposed for drag force laws and heat transfer (Tenneti *et al.* 2013). Later, Deen *et al.* (2014) proposed a correlation based on the results of Tavassoli *et al.* (2013) and Tenneti *et al.* (2013). Tenneti *et al.* (2013) simulations were carried out for a limited range of Reynolds numbers and solid volume fractions. As a result, Sun *et al.* (2015) took over this work and extended it to larger values of Reynolds numbers and solid volume fractions, and proposed a correlation that fits their PR-DNS data. The last interesting work dedicated to this topic is the simulation and analysis of Massol (2004) for monodispersed FCC arrays of spheres using AVBP code (Schönfeld and Rudgyard 1999).

The purpose of this chapter is to propose an original alternative to Lagrange extrapolation in the hydrodynamic force and heat transfer computation in order to avoid contaminated solution of the Navier–Stokes equations near the solid–fluid interface. This original method is based on Aslam extension (Aslam 2004) to extrapolate velocity, pressure and temperature from the uncontaminated fluid area to the interface. This approach aims to improve the results obtained by Lagrange extrapolation (Chadil *et al.* 2018a, c).

This chapter is structured as follows. In section 3.2, the model and numerical methods used in the VPM are briefly recalled and a description as well as a validation of Aslam extension are detailed. Thereafter, section 3.3 is devoted to setting up Aslam extension numerical parameters and estimating drag and Nusselt coefficients for uniform flows around an isolated sphere, and comparing them to those given by Lagrange extrapolation in order to assess the improvements made by Aslam extension. The same study is presented in section 3.4 for flows through fixed arrangements of mono- and bidispersed spheres. Finally, conclusions are drawn in section 3.5.

3.2. Numerical methodology

This section is a presentation of the numerical method (i.e. viscous penalty method (Vincent *et al.* 2014)) used in this chapter to simulate various particulate flows and the method used to extract from these simulations the momentum and heat transfers between the two phases (i.e. solid and fluid). This method is based on the Aslam extrapolation (Aslam 2004), which will be fully detailed, as well as its validation.

3.2.1. *Viscous penalty method*

The viscous penalty method, used in this chapter to perform particle-resolved direct numerical simulations, is a fictitious domain method where fixed staggered Cartesian grids are used to discretize both fluid and solid media. As explained by Kataoka (1986) for fluid/fluid two-phase flows and Vincent *et al.* (2014) for particle flows, the resulting model implicitly takes into account the coupling between different phases separated by resolved interfaces, i.e. larger than the mesh cell size. Given that all the configurations simulated in this chapter involve fixed particles, the motion equations then read:

$$\nabla \cdot \mathbf{u} = 0 \qquad [3.1]$$

$$\rho\left(\frac{\partial \mathbf{u}}{\partial t} + (\mathbf{u}\cdot\nabla)\,\mathbf{u}\right) = -\nabla p + \rho\mathbf{g} + \nabla\cdot\left[\mu(\nabla\mathbf{u}+\nabla^t\mathbf{u})\right] + \mathbf{F}_m \qquad [3.2]$$

where $\mathbf{u}$ is the velocity, p is the pressure, t is the time, $\mathbf{g}$ is the gravity vector, and ρ and μ are, respectively, the density and viscosity of the equivalent fluid. The source term $\mathbf{F}_m$ is used to impose a flow rate to the fluid if required.

The one-fluid model is almost identical to the classical incompressible Navier–Stokes equations, except that the local properties of the equivalent fluid (ρ and μ) depend on C. In this chapter, an arithmetic average is used for density ($\rho = C\rho_s + (1-C)\rho_f$) and a harmonic average is considered for viscosity $\left(\mu = \dfrac{\mu_s\mu_f}{C\mu_f + (1-C)\mu_s}\right)$ (Vincent *et al.* 2014).

Satisfying the incompressible and solid constraints in fluid and particles requires developing a specific model. Two penalty approaches are proposed and detailed in the next section to tackle with these constraints:

– ensuring the solid behavior in the solid zones where $C = 1$ requires a specific rheological law to be defined for the rigid fluid part without imposing the velocity, as the particle velocities are not always known *a priori* in particulate motions (particle sedimentation, fluidized beds, turbulence particle interaction). A specific model is implemented for handling the solid particle behavior in the one-fluid Navier–Stokes equations. It is based on a decomposition of the viscous stress tensor and on a penalty method that acts on the viscosity, which tends to large values in the particles (Caltagirone and Vincent 2001) to implicitly impose the solid behavior and also the coupling between fluid and solid motion. For fixed particles, the velocity of the cell containing the centroid of the particle is imposed equal to zero. The viscous penalty method propagates the zero velocity in the whole solid medium;

– following a similar walkthrough as in the work on Stokes and Navier–Stokes equations proposed by Fortin and Glowinski (1982), an augmented

Lagrangian method is applied to the unsteady Navier–Stokes equations dedicated to particulate flows. The main objective is to deal with the coupling between velocity and pressure and to satisfy the fluid and solid constraints at the same time. Starting with $\mathbf{u}^{*,0} = \mathbf{u}^n$ and $p^{*,0} = p^n$, the augmented Lagrangian solution reads while $||\nabla \cdot \mathbf{u}^{*,\mathbf{m}}|| > \epsilon$, solve

$$\begin{aligned} &(\mathbf{u}^{*,0}, p^{*,0}) = (\mathbf{u}^n, p^n) \\ &\rho^n \left(\frac{\mathbf{u}^{*,m} - \mathbf{u}^{*,0}}{\Delta t} + \mathbf{u}^{*,m-1} \cdot \nabla \mathbf{u}^{*,m} \right) - \nabla (r \nabla \cdot \mathbf{u}^{*,m}) \\ &= -\nabla p^{*,m-1} + \rho^n \mathbf{g} + \nabla \cdot [\mu^n (\nabla \mathbf{u}^{*,m} + \nabla^T \mathbf{u}^{*,m})] + \mathbf{F}_{si} + \mathbf{F}_m \\ &p^{*,m} = p^{*,m-1} - r \nabla \cdot \mathbf{u}^{*,m} \end{aligned} \quad [3.3]$$

where r is an augmented Lagrangian parameter used to impose the incompressibility constraint, m is an iterative convergence index and ϵ is a numerical threshold controlling the constraint. The augmented Lagrangian method is a kind of penalty technique: if $r \to +\infty$, the incompressibility is imposed, but solving the linear system is difficult with iterative solvers while $r \to 0$ does not act on the fluid constraint and keeps the conditioning of the matrix unchanged. Usually, a constant value of r is used, for example, equal to the average between the minimum and maximum eigenvalues of the linear system for Stokes flows (Fortin and Glowinski 1982). From numerical experiments, optimal values are found to be of the order of ρ_i and μ_i in each phase (fluid or solid) to accurately solve the motion equations in the related zone (Vincent *et al.* 2004, 2007). Algebraic improvements have also been proposed by Vincent *et al.* (2011) to automatically estimate the local values of r. In the present chapter, a constant value of r will be used.

The viscous penalty method has been validated on numerous analytical and experimental reference solutions in Caltagirone and Vincent (2001), Randrianarivelo *et al.* (2005), Vincent *et al.* (2007, 2011, 2014), de Motta *et al.* (2013) and Ozel *et al.* (2017).

3.2.2. *Drag force and heat flux computation using Aslam extension*

As mentioned in the introduction, the aim of this chapter is to propose an original method to compute the hydrodynamic force $\mathbf{F}$ and the heat flux Q_p of a particulate flow from the velocity, pressure and temperature fields extracted from PR-DNS simulations. They are given by:

$$\mathbf{F} = \oint_S \overline{\overline{\sigma}}.\mathbf{n} \; dS \quad \text{and} \quad Q_p = \oint_S -K_f \, \nabla T \cdot \mathbf{n} \; dS$$

The computation of $\mathbf{F}$ and Q_p consists of discretizing the particle surface S on a set of N elements called Lagrangian mesh (see Figure 3.1) such that:

$$\mathbf{F} \approx \sum_{l}^{N} \overline{\overline{\sigma}}_l.\mathbf{n}_l \; dS_l \quad \text{and} \quad Q_p \approx \sum_{l}^{N} -k_f \; (\nabla T)_l \cdot \mathbf{n}_l \; \Delta S_l \qquad [3.4]$$

where k_f is the thermal conductivity, $\overline{\overline{\sigma}}_l$ and $(\nabla T)_l$ are, respectively, the fluid stress tensor and the temperature gradient located at the lth element center C_l, $\mathbf{n}_l$ is the ongoing normal to the lth element and ΔS_l is its area.

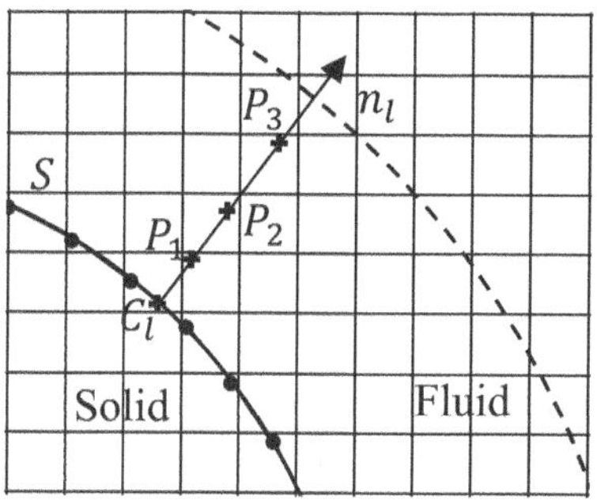

Figure 3.1. *Details of a 2D discretization of the particle surface S and extrapolation points P_i of the third-order Lagrange method used in drag force and heat flux computation*

When $\mathbf{n}_l$ and ΔS_l are easily deduced from the node coordinates of the lth element, getting the stress tensor $\overline{\overline{\sigma}}_l$ and the temperature gradient $(\nabla T)_l$ at the lth element center C_l is not straightforward when the viscous penalty method is used to simulate fluid–solid flows. Indeed, velocity, pressure and temperature values, solution of equations [3.1]–[3.2], are inaccurate in the Eulerian cells cut by the interface (Chadil *et al.* 2018a, c). To address this numerical issue, a Lagrange extrapolation coupled to a Taylor interpolation of third order was proposed to extrapolate the stress tensor (Chadil *et al.* 2018c) and the temperature gradient (Chadil *et al.* 2018a) from the fluid area far from the interface to the Lagrangian mesh (see Figure 3.1). This approach was fully detailed and validated in (Chadil *et al.* 2018a, c). These validations showed the efficiency of the method in the computation of drag force and heat flux. Indeed, the drag coefficient and the Nusselt number, both given below, were found in good agreement with many reference results found in the literature.

The force and heat transfer coefficients at the particle surface are defined as follows:

– Drag coefficient:

$$C_d = \frac{|\mathbf{F}|}{\frac{1}{2}\rho\tilde{U}^2 A_p} \quad [3.5]$$

where $\tilde{U} = |U_\infty \mathbf{e}_x - \mathbf{U}_p|$ is the relative velocity between the particle and the fluid velocity at infinity, $U_\infty \mathbf{e}_x$ is the fluid velocity in the mean flow direction far from the particle, $\mathbf{U}_p$ is the particle velocity, ν is the fluid kinematic viscosity and $A_p = \frac{\pi}{4}d^2$ is the cross-sectional area of the particle.

– Nusselt number:

$$Nu = \frac{Q_p d}{k_f\left(T_s - T_f\right)S_p} \quad [3.6]$$

where Q_p is the heat flux, k_f is the thermal conductivity and $S_p = \pi d^2$ is the area of the sphere. The temperatures of the particle and the fluid are, respectively, T_s and T_f.

By using the Lagrange extrapolation coupled to the Taylor interpolation of third order, some oscillations were observed in the distribution of both the pressure coefficient given in the spherical system (see Figure 3.2) by:

$$C_p(\theta) = \frac{p(\theta)}{\frac{1}{2}\rho U_\infty^2} \quad [3.7]$$

and the local Nusselt number

$$Nu_{loc}(\theta) = \frac{-\nabla T \cdot \mathbf{n}\, d}{(T_s - T_f)} \quad [3.8]$$

especially at a low Reynolds number (Chadil *et al.* 2018a, c). These oscillations could be an explanation for some of the differences observed between our results and literature references, even if these oscillations were observed only in a few cases among the multitude conducted in (Chadil *et al.* 2018a, c).

Another way to extrapolate the velocity, pressure and temperature from the fluid area – where they are not contaminated by the viscous penalty – to the region containing the interface, is proposed here to improve the accuracy of the current hydrodynamic force and the heat flux computations results. It is based on Aslam extensions (Aslam 2004).

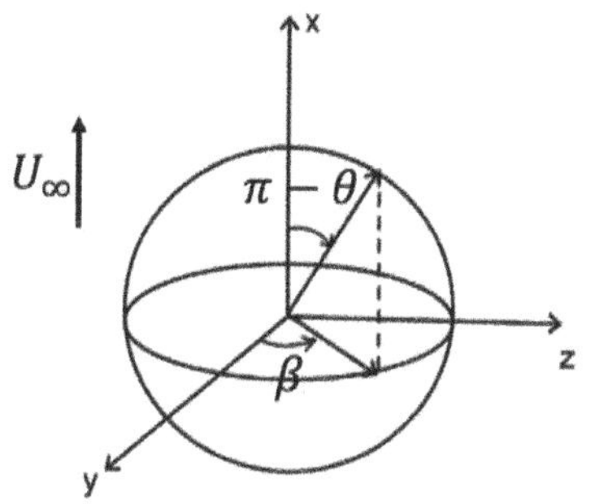

Figure 3.2. *Spherical coordinate system around a particle. The flow direction is represented by the undisturbed velocity* U_∞

3.2.2.1. *Description of Aslam extension*

Let us consider a sphere of diameter d circumscribed by a surface S, and a domain Ω divided into three sub-domains, as illustrated in Figure 3.3:

– Ω_1 the area inside the sphere of radius $(\frac{d}{2} - \Delta x)$, where Δx is the Eulerian cell mesh size;

– Ω_2 the area outside the sphere of radius $(\frac{d}{2} + \delta)$. This is the region where the velocity, pressure and temperature are not contaminated by the viscous penalty. δ is then the distance between Ω_2 and the interface;

– a band $\mathcal{B}$ bounded by the surfaces of the two spheres of radius $(\frac{d}{2} - \Delta x)$ and $(\frac{d}{2} + \delta)$, respectively. It contains the interface S. It is represented by the gray area in Figure 3.3.

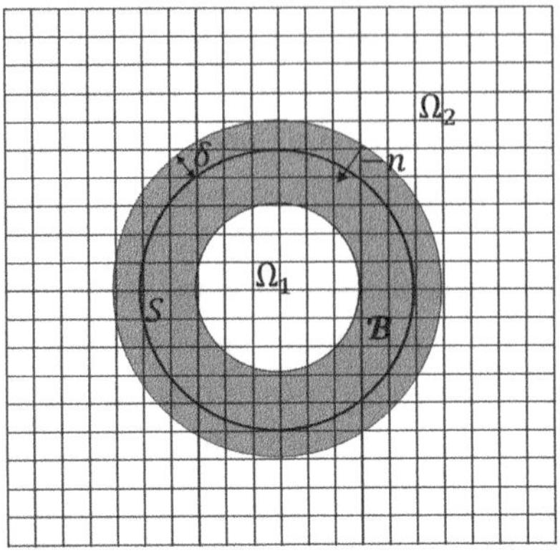

Figure 3.3. *Details of the domain subdivision for Aslam extension (Aslam 2004)*

Consider also a function g defined in Ω_2. For example, g is a velocity component or the pressure used to compute the hydrodynamic force, or the temperature used to estimate the heat flux coefficient at the particle surface. Applied to these quantities, the Aslam extension (Aslam 2004) consists then of extrapolating g from Ω_2 to $\mathcal{B}$ by solving a series of advection equations given by [3.9].

Assuming a signed distance level set function ψ is available on $\mathcal{B}$ ($\psi > 0$ inside the particle), a normal field to S inside the band can be defined as:

$$\mathbf{n} = \frac{\nabla\psi}{||\nabla\psi||}$$

In our approach and for a particle of any form, ψ is computed by a ray-casting method (Sarthou *et al.* 2009) based on the use of the Lagrangian mesh describing the particle surface S. However, for a spherical particle of diameter d and centered in $\mathbf{x_0}$, $\mathbf{n}$ and ψ are given by:

$$\forall \boldsymbol{x} \in \mathcal{B} \qquad \mathbf{n} = \frac{\boldsymbol{x} - \boldsymbol{x_0}}{||\boldsymbol{x} - \boldsymbol{x_0}||} \qquad \psi = \frac{d}{2} - ||\boldsymbol{x} - \boldsymbol{x_0}||$$

To perform an mth-order Aslam extrapolation, equations [3.9] are successively resolved for decreasing values of $k = m, m-1, ..., 1$:

$$\frac{\partial g_k}{\partial \sigma} + H\nabla g_k \cdot (-\mathbf{n}) = g_{k+1} \qquad [3.9]$$

where σ is a fictitious time, H is a Heaviside function defined by:

$$H(\boldsymbol{x}) = \begin{cases} 1 \text{ if } \boldsymbol{x} \in \mathcal{B} \\ 0 \text{ otherwise} \end{cases}$$

and g_k is the projection on the normal field of the kth-order differential of the function g:

$$\begin{aligned} g_1 &= g \\ g_2 &= \nabla g_1 \cdot (-\mathbf{n}) \\ g_3 &= \nabla g_2 \cdot (-\mathbf{n}) \\ &\vdots \\ g_m &= \nabla g_{m-1} \cdot (-\mathbf{n}) \\ g_{m+1} &= 0 \end{aligned}$$

These derivatives are computed in Ω_2 using a second-order upwind finite difference scheme:

$$(g_k)_i = \frac{3(g_{k-1})_i - 4(g_{k-1})_{i-1} + (g_{k-1})_{i-2}}{2\text{sgn}(\boldsymbol{n}_i)\Delta \boldsymbol{x}_i} \boldsymbol{n}_i \quad [3.10]$$

Using an upwind scheme rather than a central finite difference scheme, as proposed in Aslam (2004), allows the use of the same Heaviside function H for all m equations [3.9] resolved for the mth-order Aslam extension, and not a specific Heaviside function H_k for each equation, as proposed by Aslam (2004).

Equations [3.9] are resolved until a steady state is reached, using a\break first-order upwind finite difference scheme for the time derivative and a second-order TVD "SuperBee" scheme (Sweby 1984) for the space derivative. It is worth noting that when the steady state is reached in the $\mathcal{B}$ band, g verifies in $\mathcal{B}$:

$$\begin{aligned} \nabla g_m \cdot (-\mathbf{n}) &= 0 \\ \nabla g_{m-1} \cdot (-\mathbf{n}) &= g_m \\ &\vdots \\ \nabla g_2 \cdot (-\mathbf{n}) &= g_3 \\ \nabla g \cdot (-\mathbf{n}) &= g_2 \end{aligned}$$

Once the velocity and pressure fields for hydrodynamic force computation, and temperature for heat flux computation, are extrapolated in the band $\mathcal{B}$ using the mth-order Aslam extension, the stress tensor and the temperature gradient are computed in $\mathcal{B}$ and then interpolated onto the Lagrangian mesh using an mth-order Taylor interpolation, fully detailed in Chadil *et al.* (2018c). This interpolation is formulated as:

$$g(\boldsymbol{C}_l) = \sum_{|\alpha|=0}^{m-1} \frac{1}{\alpha!}\frac{\partial^\alpha g}{\partial x^\alpha}(\boldsymbol{E})(\boldsymbol{C}_l - \boldsymbol{E})^\alpha + \mathrm{O}(||\boldsymbol{C}_l - \boldsymbol{E}||^m) \qquad [3.11]$$

where $\alpha = (\alpha_1, \alpha_2, \alpha_3) \in \mathbb{N}^3$ is the sum multi-index and $\boldsymbol{E}$ denotes the nearest Eulerian point to the point $\boldsymbol{C}_l$, as illustrated in Figure 3.4 for a third-order Taylor interpolation.

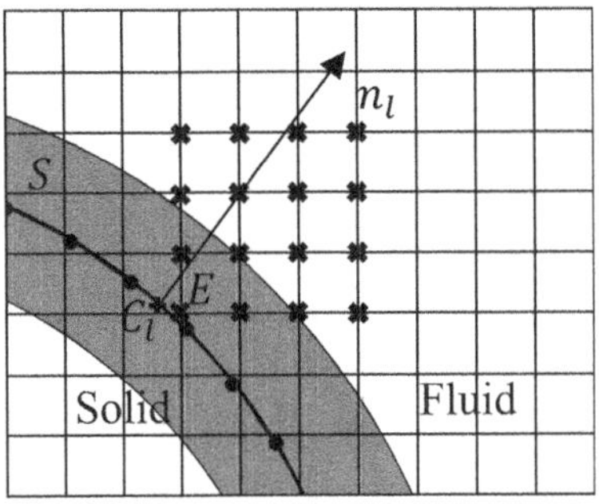

Figure 3.4. *Details of Taylor interpolation points for drag force/heat flux computation at point* C_l *of the surface Lagrangian mesh. E is the nearest fluid point to the particle/fluid interface. The crosses represent the discrete compact support of the fluid points required for a third-order Taylor interpolation*

3.2.2.2. *Validation of Aslam extension*

To validate these extensions, we used the same example as Aslam (2004), i.e. a $[-\pi, \pi]^2$ 2D domain with a particle located at the center of the domain and a function g to be extrapolated given by:

$$g(\boldsymbol{x}) = \begin{cases} \cos(\boldsymbol{x}_1)\sin(\boldsymbol{x}_2) \text{ if } \psi(\boldsymbol{x}) > 0 & \text{(i.e. inside the particle)} \\ 0 \qquad\qquad\qquad\qquad \text{otherwise} \end{cases}$$

In this example, the function g is defined inside the particle, i.e. in Ω_1. It has to be extrapolated in the band $\mathcal{B}$, as illustrated in Figure 3.5.

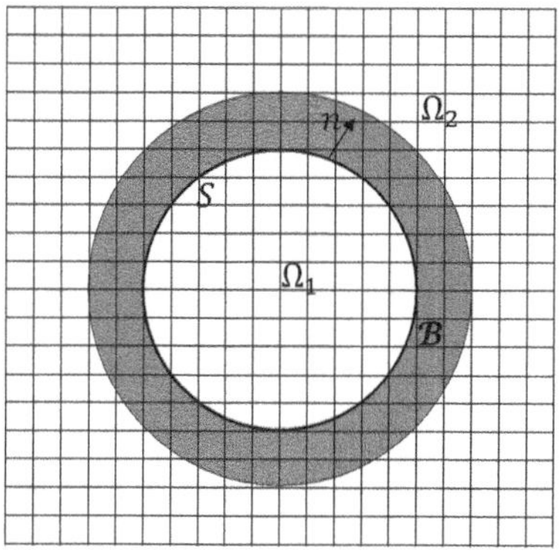

Figure 3.5. *Details of the Aslam extension validation example*

This example was extended to non-spherical particles, with an ellipse and a square. Figure 3.6 shows the contours of g inside (a) a circle, (b) an ellipse and (c) a square, which will be extrapolated in $\mathcal{B}$, i.e. the delimited white zone outside the particle, using equations [3.9].

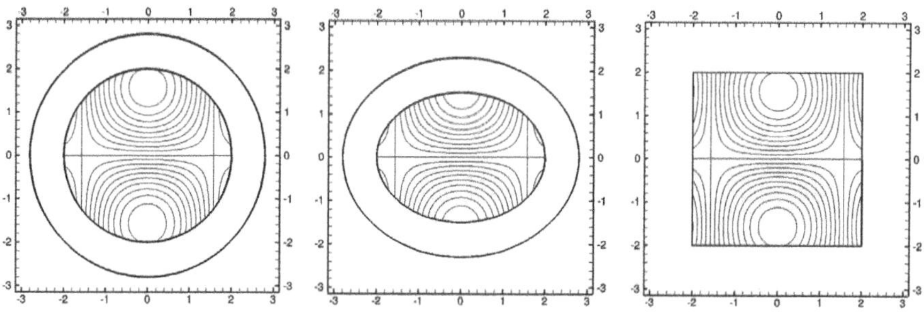

Figure 3.6. *Contour (with a 0.2 increment) of the initial conditions of g for Aslam extension from: (a) a circle, (b) an ellipse and (c) a square*

The function g is extrapolated using Aslam extension, i.e. resolving equations [3.9], for four different orders:

– Constant Aslam extension $m = 1$: this consists of resolving equation $\dfrac{\partial g}{\partial \sigma} + H\nabla g \cdot \mathbf{n} = 0$ in $\mathcal{B}$ until g reaches a steady state, i.e. $\dfrac{\partial g}{\partial \sigma} = 0$, i.e. $\dfrac{\partial g}{\partial \mathbf{n}} = \nabla g \cdot \mathbf{n} = 0$, which means that extrapolated g is constant in the normal direction to the particle surface in the extrapolation region $\mathcal{B}$, as illustrated in Figure 3.7.

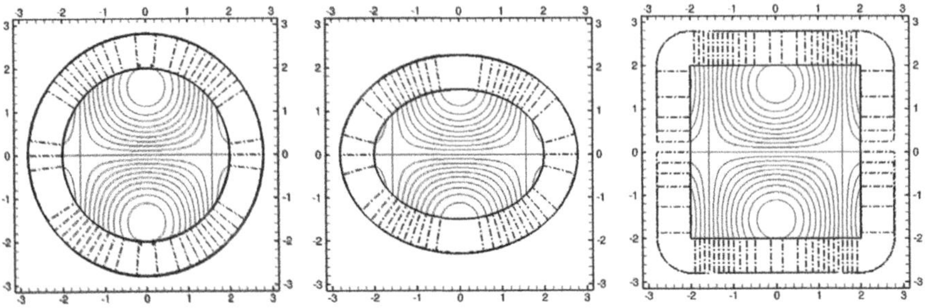

(a) (b) (c)

Figure 3.7. *Contour (with a 0.2 increment) of the function g extrapolated using constant Aslam extension ($m = 1$) from: (a) a circle, (b) an ellipse and (c) a square*

– Linear Aslam extension $m = 2$: this consists of extrapolating the first normal derivative $g_2 = \dfrac{\partial g}{\partial \mathbf{n}}$ from Ω_1, where it is computed from g using [3.10], to the band $\mathcal{B}$ using constant Aslam extension. Then, g is extrapolated from Ω_1 to $\mathcal{B}$ by resolving equation $\dfrac{\partial g}{\partial \sigma} + H\nabla g \cdot \mathbf{n} = g_2$ until g reaches a steady state. An illustration of such a function is given in Figure 3.8, where we can observe that g is no more constant in the normal direction to the particle surface in $\mathcal{B}$ ($\nabla g \cdot \mathbf{n} = g_2$).

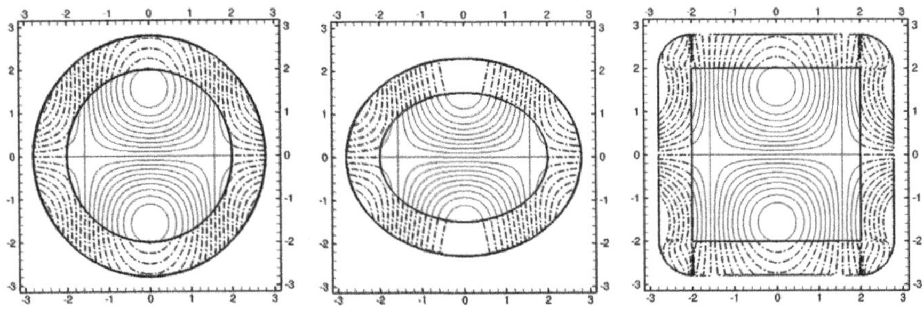

(a) (b) (c)

Figure 3.8. *Contour (with a 0.2 increment) of the function g extrapolated using linear Aslam extension ($m = 2$) from: (a) a circle, (b) an ellipse and (c) a square*

– Quadratic Aslam extension $m = 3$: this consists of extrapolating the second normal derivative $g_3 = \frac{\partial g_2}{\partial \mathbf{n}} = \frac{\partial^2 g}{\partial \mathbf{n}^2}$ from Ω_1,where it is computed from g_2 using [3.10], to the band $\mathcal{B}$ using constant Aslam extension. Then, g_2 is extrapolated using linear Aslam extension with g_3 as the second member of the extrapolation equation. And finally, g is extrapolated by resolving in $\mathcal{B}$ $\frac{\partial g}{\partial \sigma} + H\nabla g \cdot \mathbf{n} = g_2$ until g reaches a steady state. In Figure 3.9, the extrapolated part of g seems to converge towards the initial function g. Note that the incoherent values of the extrapolated g in the corner region of the square case are the consequence of the fact that the normal is unknown in this region.

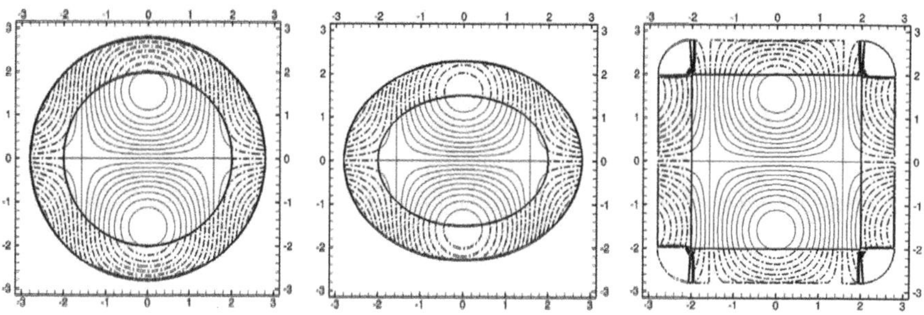

(a) (b) (c)

Figure 3.9. *Contour (with a 0.2 increment) of the function g extrapolated using quadratic Aslam extension ($m = 3$) from: (a) a circle, (b) an ellipse and (c) a square*

– Cubic Aslam extension $m = 4$: this consists of extrapolating the third normal derivative $g_4 = \frac{\partial g_3}{\partial \mathbf{n}}$ from Ω_1, to the band $\mathcal{B}$ using constant Aslam extension. In a second step, g_3 is obtained using linear Aslam extension with g_4 as a second member of the extrapolation equation. Then, g_2 is extrapolated using quadratic Aslam extension with g_3 as the second member of the extrapolation equation. Finally, g is approximated by resolving $\frac{\partial g}{\partial \sigma} + H\nabla g \cdot \mathbf{n} = g_2$ until g reaches a steady state. The results obtained with this extrapolation are illustrated in Figure 3.10.

A convergence study was conducted for these four Aslam extension orders using the relative error, in the band $\mathcal{B}$, between exact g values and those obtained with Aslam extrapolation procedures. It can be observed in

Figure 3.11 that these errors are always small (from 0.1 to 10^{-8}%) and that the expected orders are found for extensions of the circle and the ellipse. However, in the case of an extension from a square surface, the order could not be reached because of the inaccurate values in the region near the square corner, where the normal to the surface is not defined.

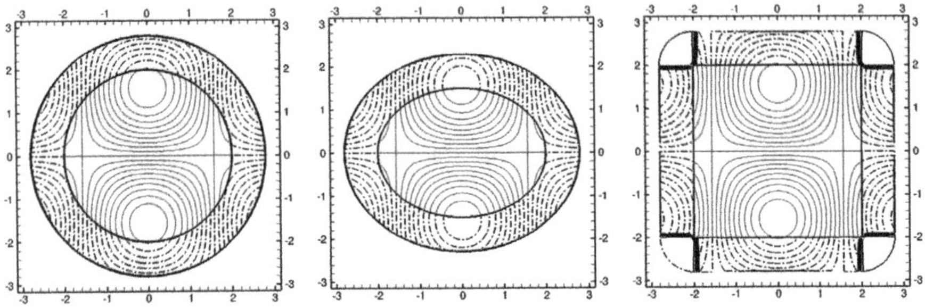

(a) (b) (c)

Figure 3.10. *Contour (with a 0.2 increment) of the function g extrapolated using cubic Aslam extension ($m = 4$) from: (a) a circle, (b) an ellipse and (c) a square*

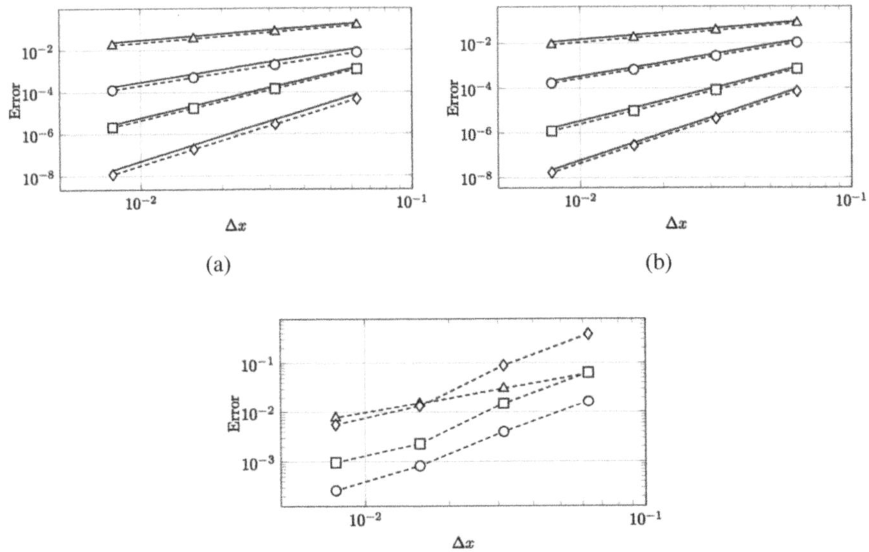

(a) (b) (c)

Figure 3.11. *Convergence orders in two dimensions for Aslam extension from: (a) a circle, (b) an ellipse and (c) a square. (△) $m = 1$, (○) $m = 2$, (□) $m = 3$ and (◇) $m = 4$*

The exact same study was conducted for three-dimensional cases. Indeed, a $[-\pi, \pi]^3$ 3D domain with a particle (sphere, ellipsoid, cube) located at the center of the domain and a function g to be extrapolated are considered here to validate Aslam extension in 3D. Now, the function g is defined by:

$$g(\boldsymbol{x}) = \begin{cases} \cos(\boldsymbol{x}_1)\sin(\boldsymbol{x}_2)\cos(\boldsymbol{x}_3) \text{ if } \psi(\boldsymbol{x}) > 0 & \text{(i.e. inside the particle)} \\ 0 & \text{otherwise} \end{cases}$$

Figure 3.12 shows that, similar to the 2D cases, the expected extension orders are found for the sphere and the ellipsoid, even if the relative errors are two orders of magnitude greater than in 2D cases. As for the 2D case, the non-definition of the normal to the cube in the vicinity of its corners leads to inaccurate values of the extrapolated function in the region near these corners. This explains the error behavior that makes the extension order computation impossible in this case.

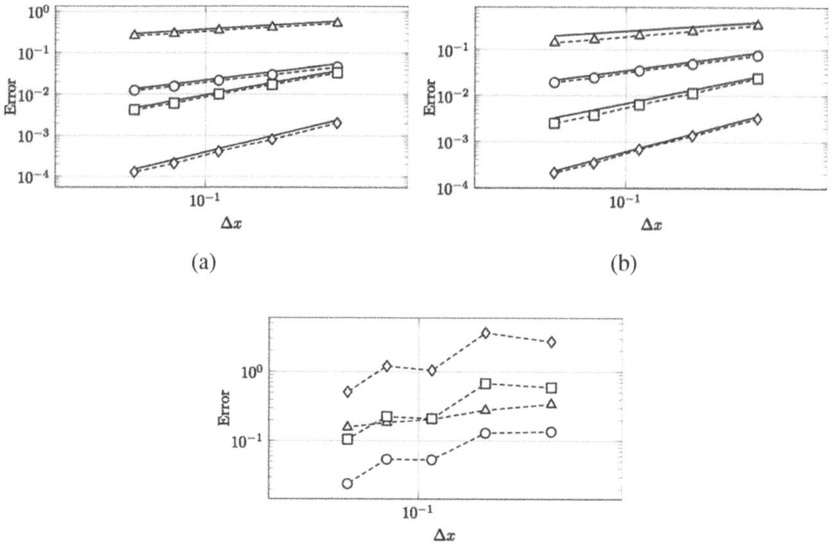

Figure 3.12. *Convergence orders in three dimensions for Aslam extension from: (a) a sphere, (b) an ellipsoid and (c) a cube. (△) $m = 1$, (○) $m = 2$, (□) $m = 3$ and (◇) $m = 4$*

3.3. Isolated stationary sphere passed by a uniform flow

The aim of this chapter is to assess how Aslam extension could improve the computation of the hydrodynamic force and the heat flux compared to the results obtained using Lagrange extrapolation. To do that, we conducted the same simulations carried out in Chadil *et al.* (2018a, c) to validate the force and heat transfer computation using Lagrange extrapolation. The first case considered here is a forced convection by uniform flow past a hot stationary sphere, illustrated in Figure 3.13. All the details and description of this problem are reported in Chadil *et al.* (2018a, c).

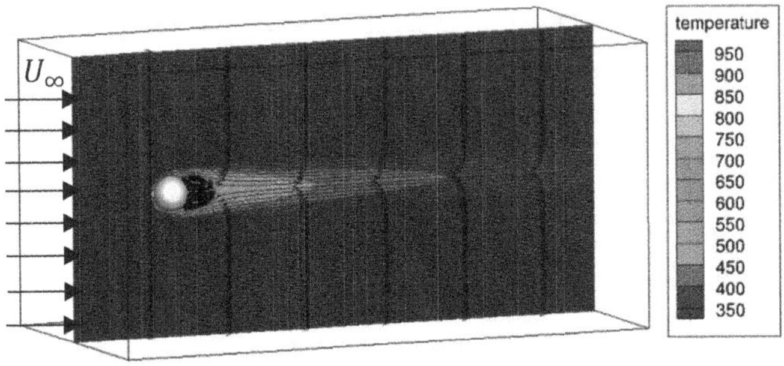

Figure 3.13. *Streamlines and temperature field for a uniform flow past a sphere at* $Re = 100$. *For a color version of this figure, see www.iste.co.uk/prudhomme/fluid1.zip*

3.3.1. *Drag force computation*

Following the steps of the study of the drag force computation using Lagrange extrapolation (Chadil *et al.* 2018c), we will first determine how far we have to go from the particle to reach uncontaminated velocity and pressure of the fluid to extrapolate them to the interface, in order to get accurate values of drag force. The effect of Aslam extrapolation order will be studied at the same time on these values. Moreover, some perturbations were observed in the local pressure distribution on the particle surface in Chadil *et al.* (2018c). We will see how Aslam extension could prevent these oscillations. Finally, we will compare the drag coefficient values computed with Lagrange extrapolation (Chadil *et al.* 2018c) with those computed with Aslam extension.

3.3.1.1. *Effect of the extrapolation distance* δ

One of the main drag force computation parameters using Lagrange extrapolation, studied in Chadil *et al.* (2018c), is the distance (δ) between the particle and the first point of extrapolation $\mathbf{P}_1$ (see Figure 3.1). Indeed, as reported in Chadil *et al.* (2018c) and as illustrated in Figure 3.14a at Stokes regime and in Figure 3.14c at $Re = 100$, the drag force values, computed using Lagrange extrapolation, are inaccurate for $\delta \leq 1$. This is due to the contaminated values of velocity and pressure induced by the viscous penalty method (Chadil *et al.* 2019) in the Eulerian cells cut by the interface. The study of the effect of this distance δ (illustrated in Figure 3.3) is conducted for drag force computation using Aslam extension as well. Figure 3.14b and 3.14d shows that, in order to get an accurate drag force using Aslam extensions, a third order is required. Therefore, for the rest of this chapter, a third-order Aslam extension coupled with a third-order Taylor interpolation will be used in the drag force computation. We can also observe in these figures that the distance to go far from the interface is reduced to $\delta = 0.5$ with the Aslam approach. One possible explanation of this improvement is that it is actually the pressure and the velocity that are extrapolated by Aslam extension in the drag force computation, rather than the stress tensor components, i.e. the gradients of these variables, as is the case when the Lagrange extrapolation is used. Indeed, the stress tensor is computed using centered difference schemes that use contaminated values of velocity and pressure for $\delta \leq 1$ (and even solid values for $\delta = 0$). On the other hand, extrapolating directly the velocity components and the pressure using Aslam extension and then computing the stress tensor from these extrapolated values reduces the distance to which we have to move away from the particle to reach accurate drag force values.

3.3.1.2. *Pressure coefficient*

Analysis of local pressure profiles in Chadil *et al.* (2018c) was used to better understand the behavior of the pressure on the sphere, depending on the Reynolds number. The local pressure coefficients are defined by equation [3.7] in a spherical coordinate system (see Figure 3.2).

The pressure coefficient distribution according to θ was compared in Chadil *et al.* (2018c) to some available body-fitted simulation results obtained by Clair *et al.* (1970), Dennis and Walker (1971), Magnaudet *et al.* (1995) and Massol (2004) at different Reynolds numbers. The results are presented in Figure 3.15a, 3.15c and 3.15e for $Re = 1, 10, 100$, respectively.

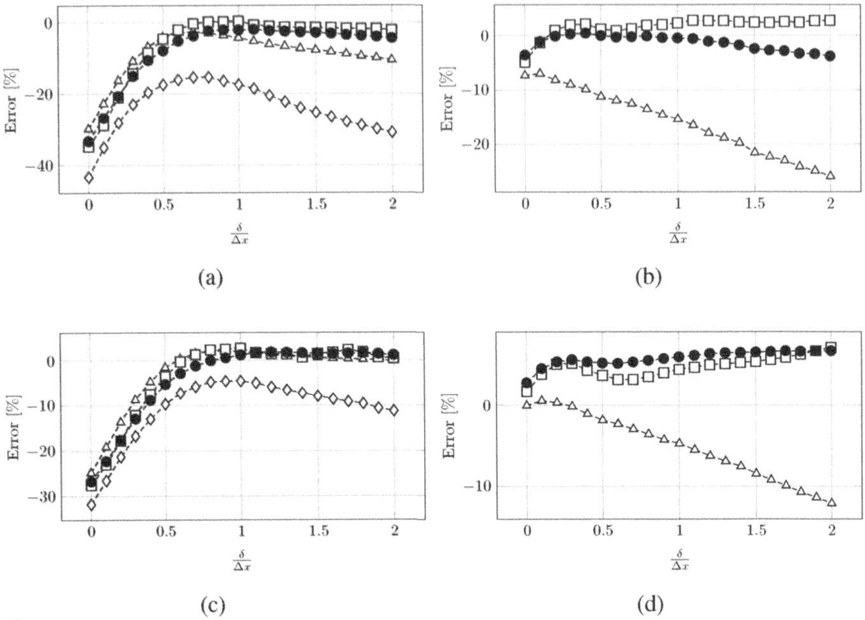

Figure 3.14. *Drag force relative error (%) for the uniform flow past a sphere in Stokes regime (a) for Lagrange extrapolation and (b) for Aslam extrapolation and at $Re = 100$ for (c) Lagrange extrapolation and (d) Aslam extrapolation. Different extrapolation orders are considered: (◇) first, (△) second, (●) third and (□) fourth orders. The distance between the first Eulerian point used to extrapolate forces and the particle surface is δ*

It is well known that, for this range of Reynolds numbers, the flow is symmetric with respect to its direction, as illustrated in Figure 3.16. Therefore, the pressure coefficient distribution must be the same for all the polar angle (β) plans, translating by that the flow symmetry. This expected behavior was mostly observed when Lagrange extrapolation was used in the drag force computation Chadil *et al.* (2018c) (see Figure 3.15a, 3.15c and 3.15e). However, some local pressure values diverge from the expected ones, especially for low Reynolds numbers ($Re = 1, 10$), as illustrated in Figure 3.15a and 3.15c.

The use of Aslam extension in the drag force computation seems to reduce the pressure oscillations, which is a good feature of the method, although only few pressure points exhibited values far from expected ones when Lagrange extrapolation is used. Indeed, the same distribution of the pressure coefficient is observed for all β plans in Figure 3.15b, 3.15d and 3.15f.

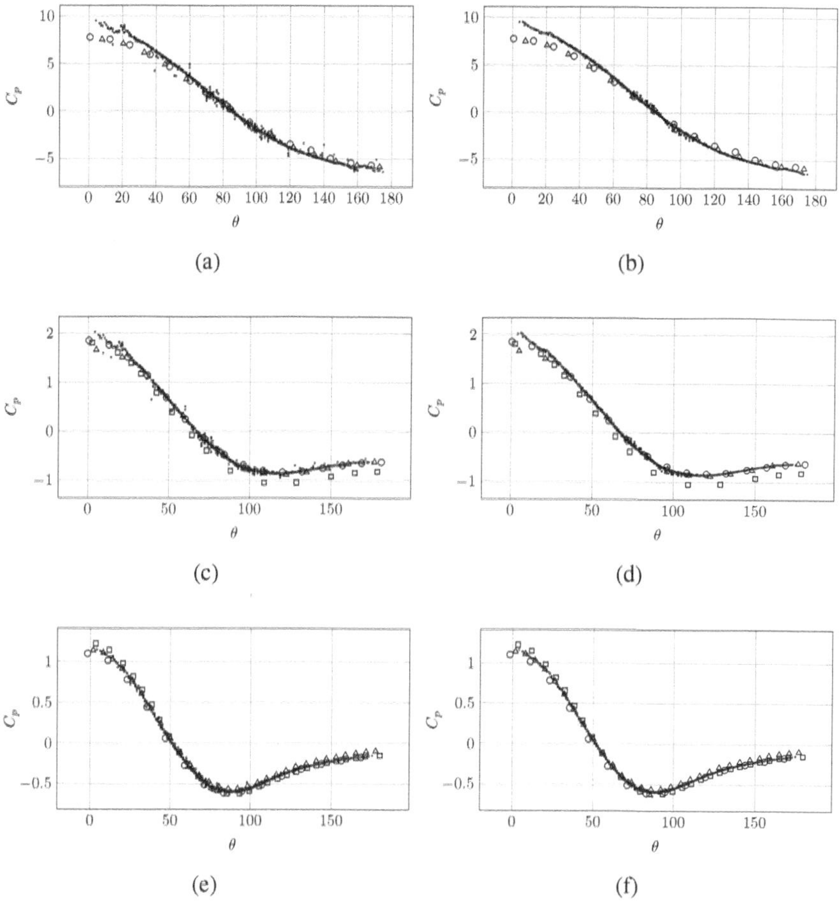

Figure 3.15. *Pressure coefficient for a uniform flow past an isolated sphere at $Re = 1$: (a) for Lagrange extrapolation, (b) for Aslam extrapolation, $Re = 10$, (c) for Lagrange extrapolation, (d) for Aslam extrapolation, $Re = 100$, (e) for Lagrange extrapolation and (f) for Aslam extrapolation. (◇) Dennis and Walker (1971), (△) Magnaudet* et al. *(1995), (○) Clair* et al. *(1970), (□) Massol (2004) and (.) present work*

3.3.1.3. *Drag coefficient*

The last step in the study of drag force computation, using a third-order Aslam extension, is devoted to the drag coefficient given by equation [3.5] and extracted from PR-DNS. It is compared to the Schiller and Naumann (1933) correlation for a wide range of Reynolds numbers ($Re = 0.1 - 290$). This comparison was conducted in Chadil *et al.* (2018c) when the drag force was computed using Lagrange extrapolation.

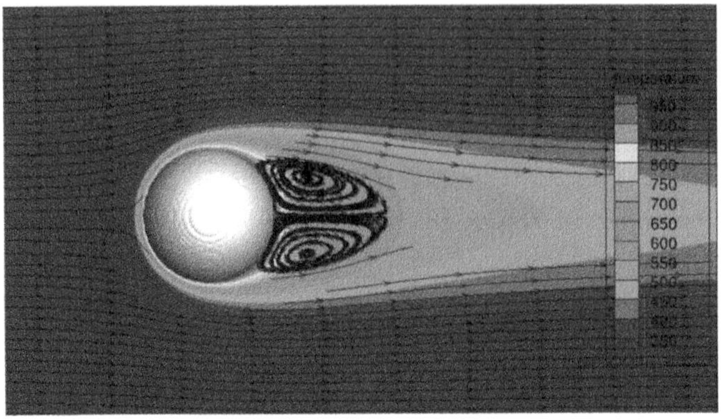

Figure 3.16. *Streamlines and temperature field for a uniform flow past a sphere at* $Re = 100$. *For a color version of this figure, see www.iste.co.uk/prudhomme/fluid1.zip*

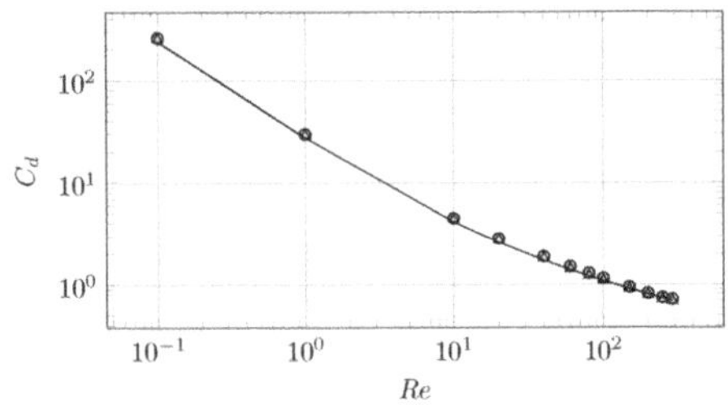

Figure 3.17. *Drag coefficient for uniform flow past sphere at different Reynolds numbers: (—) Schiller and Naumann (1933), (○) present work using Aslam extension and (△) using Lagrange extrapolation*

Figure 3.17 shows the good agreement of our results with Schiller and Naumann (1933). It also shows, at least for this particular case, that the drag force computed using Aslam extension is almost the same as the one computed using Lagrange extrapolation. Indeed, as has been discussed earlier, Aslam extension improvement was the reduction of the distance δ to go away from the particle to get accurate drag force, and the correction of the few inaccurate values of the pressure on the particle surface. In the case of an isolated sphere, the distance δ does not affect the outcome of the drag force as soon as the accurate value is found, regardless of the force calculation method used (Lagrange extrapolation or Aslam extension). On the contrary, in the case of arrangement of spheres, which will be discussed in the next section, the distance δ will be an important parameter, especially for high solid volume fractions. Moreover, the number of inaccurate pressure values in the pressure coefficient distribution is not important enough to really influence the total drag force in the present case of an isolated sphere.

3.3.2. *Heat transfer computation*

The effect of choosing δ is now considered with respect to the heat transfer computation. This effect is estimated using the local Nusselt number computed using a third-order Aslam extension (Aslam 2004) coupled to a third-order Taylor interpolation, followed by a comparison of this local Nusselt number distribution to the one computed using Lagrange extrapolation (Chadil *et al.* 2018a). Once the parameter δ is set, the global Nusselt number for the problem of a uniform flow past a hot sphere is compared to the correlations given by Ranz and Marshall (1952), Whitaker (1972) and Feng and Michaelides (2000).

3.3.2.1. *Effect of the extrapolation distance* δ

To study the δ effect, the local Nusselt number distribution was compared to the one provided by Massol (2004) thanks to the body-fitted PR-DNS results at $Re = 100$. As previously explained for the pressure coefficients, this distribution is also expected to be symmetric with respect to the flow direction, as illustrated in Figure 3.16.

The local Nusselt number computed for $\delta = 0$ does not reflect the flow symmetry, either for the one computed using Lagrange extrapolation (see Figure 3.18a) or for the one computed using Aslam extension (see Figure 3.18b), even if the latest significantly improved the result with a lower amplitude of Nusselt oscillations. Indeed, the error between our result (Nu_{loc}) and Massol's $(Nu_{locMassol})$ given by $\frac{\sum_{i=1}^{N} |(Nu_{loc})_i - (Nu_{locMassol})_i|}{\sum_{i=1}^{N} |(Nu_{locMassol})_i|}$ (illustrated in Figure 3.19) is about 25% for Lagrange extrapolation and less than 15% for the Aslam extension. This error decreases as the distance δ increases until being lower than 10% for $\delta \geq 1$ in the case of Lagrange extrapolation use, and about 5% in the case of Aslam extension use for $\delta = 1$, as illustrated in Figure 3.19. Moreover, Figure 3.18c and 3.18d shows the distribution of the local Nusselt number for $\delta = 1$ with, respectively, Lagrange and Aslam extrapolations, and they also show that Aslam extension improves the heat transfer computation with an almost symmetrical distribution of the local Nusselt number for $\delta = 1$. Therefore, the extrapolation distance will always be $\delta = 1$ in the rest of this chapter.

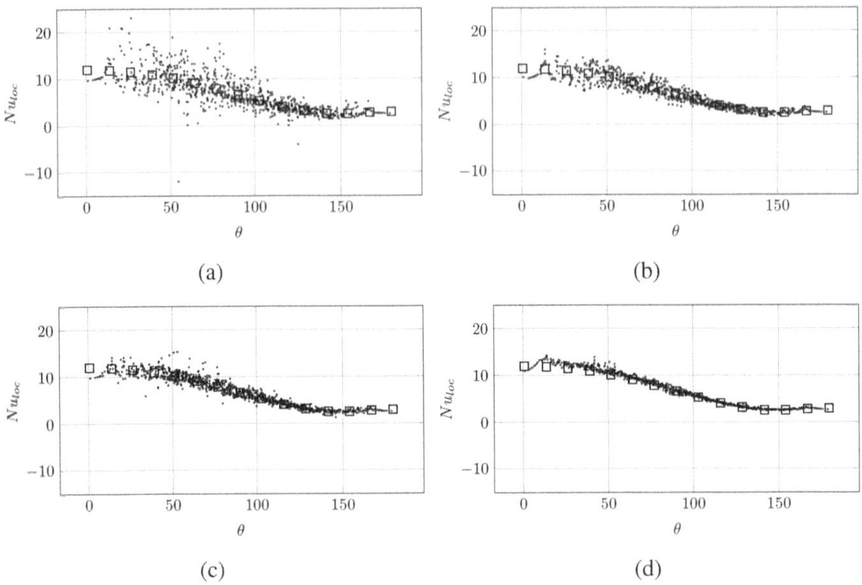

Figure 3.18. *Local Nusselt coefficient for a uniform flow past an isolated sphere at $Re = 100$, as a function of the azimuthal angle θ at $\delta = 0$: (a) for Lagrange extrapolation, (b) for Aslam extrapolation, $\delta = 1$, (c) for Lagrange extrapolation and (d) for Aslam extrapolation. (.) present work, (□) Massol (2004)*

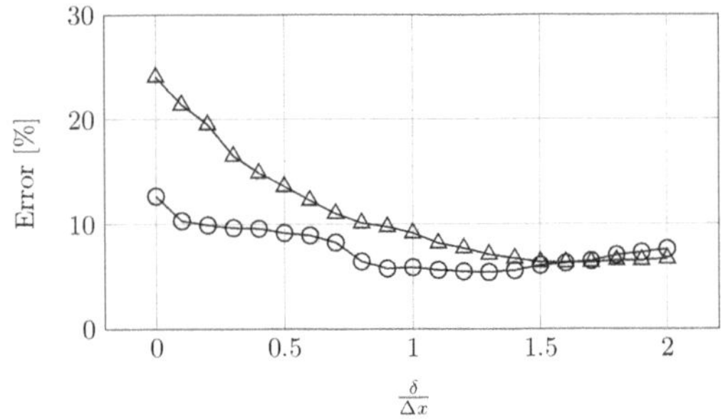

Figure 3.19. *Local Nusselt number relative error compared to Massol's result (Massol 2004) as a function of the extrapolation distance δ: (○) using Aslam extension and (△) using Lagrange extrapolation*

3.3.2.2. *Nusselt number*

After setting up the numerical parameter for the heat transfer computation, the Nusselt number obtained, when considering a uniform flow past a hot fixed sphere, is computed using Aslam extension and compared to the values computed using Lagrange extrapolation (Chadil *et al.* 2018a) and to some of the existing correlations of the literature.

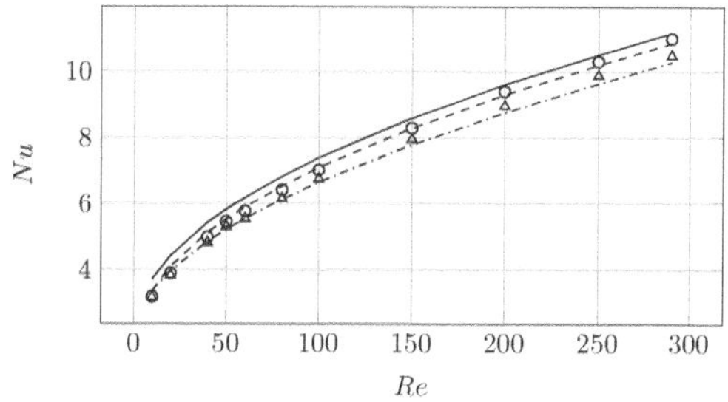

Figure 3.20. *Nusselt coefficient for the uniform flow past a hot sphere at various Reynolds numbers: (——) Ranz and Marshall (1952), (- - -) Feng and Michaelides (2000), (-·-·) Whitaker (1972), (○) present work using Aslam extension and (△) using Lagrange extrapolation*

Figure 3.20 shows the good agreement of our results to correlations, with clear improvement when Aslam extension is used to compute the Nusselt number. Indeed, unlike the drag force computation, and as discussed in the previous section, the Aslam extension has an effect on the obtained local Nusselt number values. This naturally improves the global Nusselt number that converges towards the well-known Ranz and Marshall correlation (Ranz and Marshall 1952) (see Figure 3.20), which is expected to be the reference law used in the literature.

3.4. Face-centered cubic arrangement of stationary sphere passed by a uniform flow

As explained in section 3.3.1.3, Aslam extension does not improve the drag force computation of an isolated sphere. Indeed, Aslam extension reduced the distance δ to go far from the particle to reach accurate drag value, going from $\delta = 1$ for Lagrange to $\delta = 0.5$ for Aslam. Moreover, for an isolated sphere, this distance does not affect the computed value of the drag as long as an accurate value is reached, regardless of the extrapolation technique used (see Figure 3.14). However, it is obvious that for an assembly of spheres, the distance δ would affect the drag force computed on each sphere as the distance between the spheres reduces when the solid volume fraction increases. Two configurations of sphere assemblies – mono- and bidispersed arrangements of spheres – are studied in this section, following the work done in Chadil *et al.* (2018a, c) where Lagrange extrapolation was used in the drag force and heat transfer computation in order to compare the two extrapolation method outcomes.

3.4.1. *Monodispersed face-centered cubic periodic arrangement of spheres*

The first case considered here is a uniform flow past a face-centered cubic (FCC) array of spheres. We wish to study whether or not Aslam extension improves the drag force and the heat transfer computations when a sphere is surrounded by other particles, in comparison to those computed using Lagrange extrapolation. FCC arrangement of spheres consists of a cube where three spheres are placed on the face centers, and one sphere is located on the vertices with periodic boundary conditions, as illustrated in Figure 3.21. This case is detailed fully in Chadil *et al.* (2018a, c).

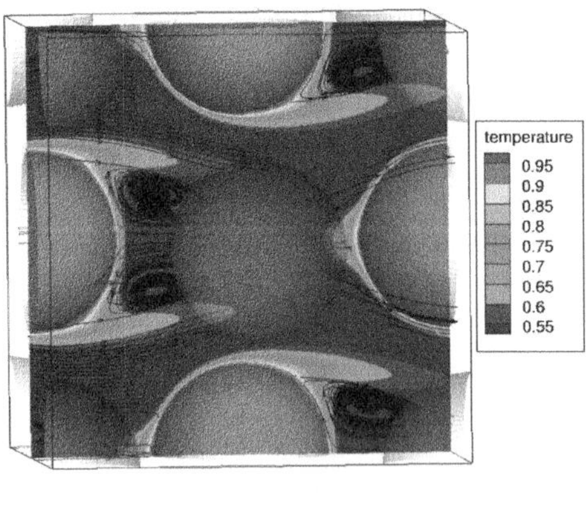

Figure 3.21. *Streamlines and temperature field for a steady flow along the x-axis of a face-centered cubic array of spheres at* $Re = 300$, $\alpha_d = 0.15$. *For a color version of this figure, see www.iste.co.uk/prudhomme/fluid1.zip*

3.4.1.1. *Drag force*

When the steady-state flow is reached at the desired Reynolds number, as explained in Chadil *et al.* (2018c), the mean non-dimensional drag force F for all the particles is then deduced from the drag forces computed over each particle by:

$$F = \frac{|\langle \mathbf{F}^D \rangle|}{3\pi\mu d(1-\alpha_d)|\langle \mathbf{u}_f \rangle|} \quad [3.12]$$

with $\langle \mathbf{F}^D \rangle = \frac{1}{N_p}\sum_i^{N_p} \mathbf{F}_i^D$ and $\mathbf{F}_i^D$ being the drag force computed over the ith particle.

The non-dimensional drag force F normalized by the isolated sphere non-dimensional drag force (given by Schiller and Naumann (1933)) $F_s = \frac{1+0.15Re^{0.687}}{(1-\alpha_d)^2}$, denoted as $\frac{F}{F_s}$, is compared to Massol's results (Massol 2004) extracted from body-fitted simulations, and taken here as reference results, and to the drag force results computed using Lagrange extrapolation, in order to assess the Aslam extension effect on the drag force computation.

Figure 3.22 shows the non-dimensional normalized drag force $\frac{F}{F_s}$ computed using Aslam extension, the one computed using Lagrange extrapolation and finally Massol's results (Massol 2004) used here as a reference to illustrate the possible improvement brought by the Aslam extension being applied to the drag force computation. We can observe that for $Re = 10, 50, 100$, the drag force computed using Aslam extension converges towards the reference body-fitted values provided by Massol (2004). Aslam simulations match Massol's results better than the values computed using Lagrange extrapolation. This clearly shows the positive effect of the reduction, by Aslam extension, of the extrapolation distance δ on the drag force computation. It is worth noting that Massol simulations were under-resolved for $Re = 300$, which could explain the discrepancies in our results.

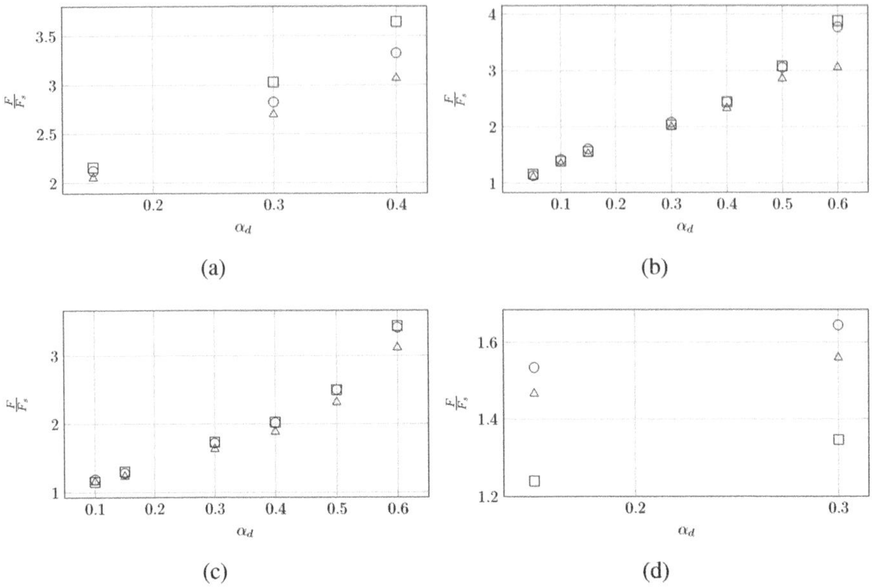

Figure 3.22. *Drag force for a uniform flow past an FCC, normalized by Schiller and Naumann (1933) drag force for a uniform flow past an isolated sphere. Results are presented as a function of the solid volume fraction α_d and Reynolds number: (a) $Re = 10$, (b) $Re = 50$, (c) $Re = 100$, (d) $Re = 300$: (□) Massol (2004), (○) present work using Aslam extension and (△) present work using Lagrange extrapolation*

3.4.1.2. *Nusselt coefficient*

Although Aslam extension improvement was observed for an isolated sphere on the Nusselt number results (unlike for the drag force), a study of its effect is conducted here for an FCC array of spheres, in order to test its efficiency for the heat transfer computation in an assembly of spheres. The global Nusselt coefficient for a uniform flow past an FCC array of spheres is deduced from the Nusselt computed on each sphere by:

$$\langle Nu \rangle = \frac{1}{4} \sum_{i}^{N_p} Nu_i \qquad [3.13]$$

where Nu_i is the Nusselt number computed over the ith particle.

Figure 3.23 shows the normalized global Nusselt coefficient $\frac{Nu}{Nu_s}$ for a uniform flow past an FCC array of spheres, computed, on the one hand, using

Aslam extension and, on the other hand, Lagrange extrapolation. It also shows $\frac{Nu}{Nu_s}$ extracted from Massol's body-fitted simulations (Massol 2004). We can observe that Aslam extension improves the heat transfer computation for a packed sphere even if it seems to overestimate it with respect to Massol's results for high solid volume fraction.

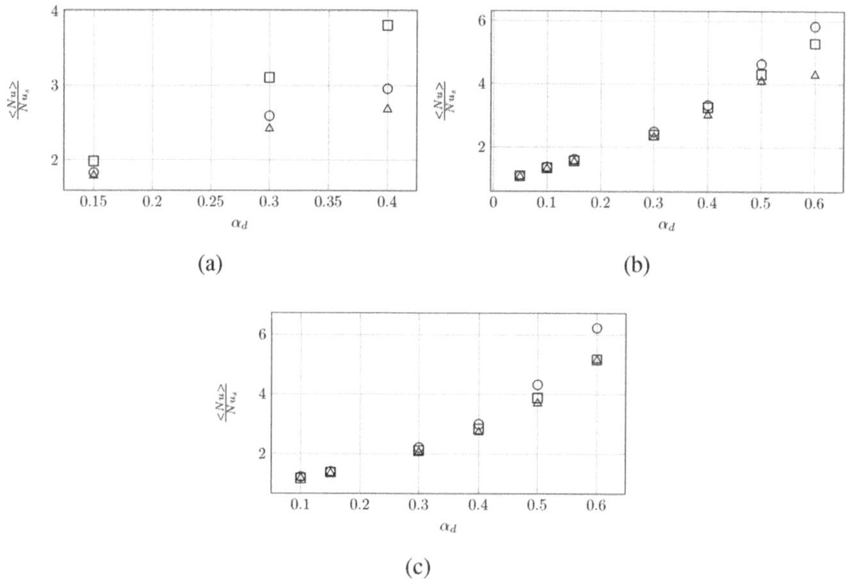

Figure 3.23. *Global Nusselt coefficient for a uniform flow past an FCC, normalized by Ranz and Marshall (1952) Nusselt coefficient for a uniform flow past an isolated sphere. Evolution as a function of the solid volume fraction α_d at Reynolds numbers (a) $Re = 10$, (b) $Re = 50$ and (c) $Re = 100$: (□) Massol (2004), (○) present work using Aslam extension and (△) present work using Lagrange extrapolation*

3.4.2. *Bidisperse face-centered cubic periodic arrangement of spheres*

The last case investigated in this chapter to assess the possible improvement brought when Aslam extension is used for the drag force computation is a bidisperse face-centered cubic periodic arrangement for two species of spheres: the larger particles are distributed in the same configuration as the one previously presented for monodisperse arrangements, while the smaller particles are located at the center of the vertices and at the center of the cubic simulation domain, as illustrated in Figure 3.24. The parameters to set up this case are detailed fully in Chadil *et al.* (2018c).

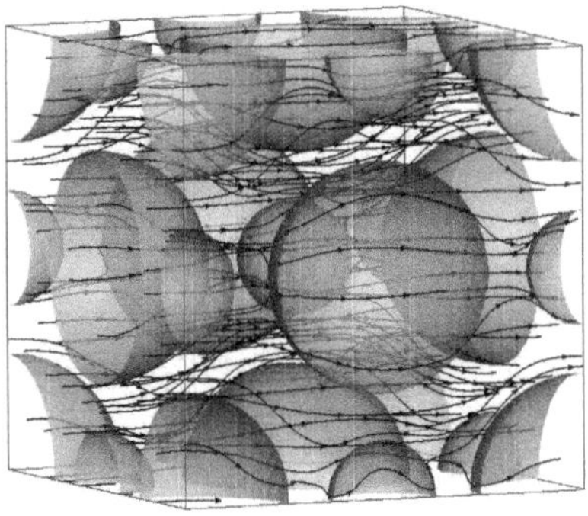

Figure 3.24. *FCC/FCC bidisperse arrangement of spheres for $Re = 50$ and $\alpha_d = 0.3$ – the particle shape is plotted in blue, and the black lines are the streamlines. For a color version of this figure, see www.iste.co.uk/prudhomme/fluid1.zip*

The two species are characterized by the following dimensionless parameters:

$$x_i = \frac{\alpha_i}{\alpha_d} \quad , \quad y_i = \frac{d_i}{d_s} \qquad [3.14]$$

where d_i and α_i are the particle diameter and the solid volume fraction of the specie i, respectively. The Sauter mean diameter d_s is given by:

$$d_s = \left[\sum_i^2 \frac{x_i}{d_i} \right]^{-1}$$

Figure 3.25 shows the non-dimensional drag force F for a uniform flow past an FCC/FCC packed sphere, computed using Aslam extension, as well as the one computed using Lagrange extrapolation (Chadil *et al.* 2018c). The results are compared to the body-fitted simulations of Massol in terms of drag force, taken as reference values. It can be observed that drag forces computed using Aslam extension are closer to Massol's results than those computed using Lagrange extrapolation, and that for both species y_1 and y_2 at $Re = 50, 100$.

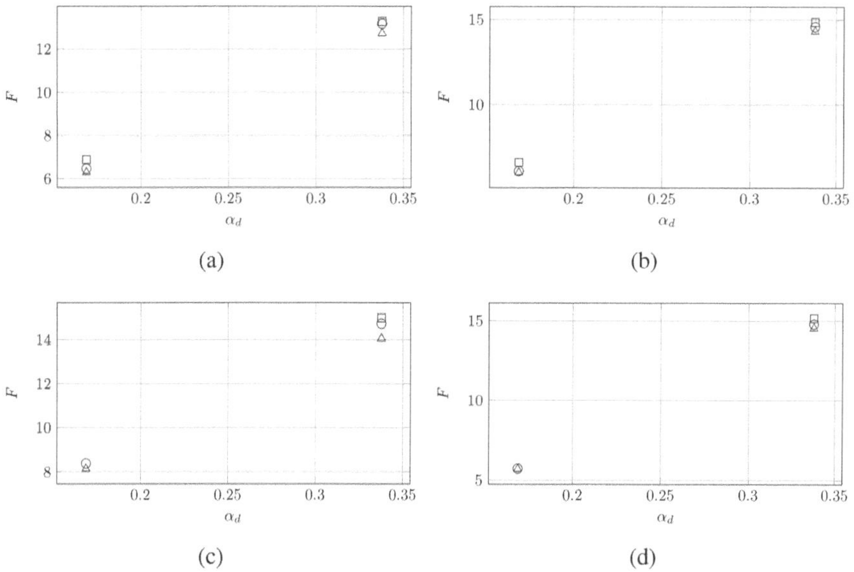

Figure 3.25. *Non-dimensional drag force F for a uniform flow past an FCC/FCC packed sphere as a function of the solid volume fraction α_d (a): $(Re = 50, y1)$, (b): $(Re = 50, y2)$, (c): $(Re = 100, y1)$, (d): $(Re = 100, y2)$: (□) Massol (2004), (○) present work using Aslam extension and (△) present work using Lagrange extrapolation*

3.5. Conclusion

A new method is proposed to compute momentum and heat transfer in particle-resolved direct numerical simulations. It is based on Aslam extension (Aslam 2004) coupled with Taylor interpolation, both of third order. This method was first validated without flow resolution for various particle shapes (spherical, ellipsoidal and cubical particles), and a numerical study has been conducted in both two and three dimensions. The expected convergence orders were obtained for spherical and ellipsoidal particles, whereas the existence of non-convex regions in the cubical particles generates computation errors that made the convergence study impossible.

This method is constructed to be a better alternative to the approach proposed for drag force and heat transfer estimates in Chadil *et al.* (2018a, c), respectively. This improvement was shown by comparing the results of both techniques applied to multiple test cases. First, in a uniform flow past an isolated hot sphere, the use of Aslam extension, on the one hand, significantly reduces the extrapolation distance (i.e. the distance needed to no longer be affected by the contaminated value of pressure, velocity and temperature caused by the averaging of physical characteristics of the mixture in the Eulerian cells containing the particle/fluid interface). On the other hand, it corrects some inaccurate local pressure values on the sphere observed in the Lagrange extrapolation results (Chadil *et al.* 2018c). The effect of Aslam extension on the drag and Nusselt coefficients was more significant in more complex flows as mono- and bidisperse faced-centered cubic arrangement of spheres.

The general conclusion that we have obtained is that drag force and heat transfer calculations using Aslam extensions compare very well with body-fitted simulations of Massol (2004) better than the computation using Lagrange extrapolations.

Future works will be devoted to applying our hydrodynamic force and heat transfer calculation method to random arrangements of spherical and ellipsoidal particles, and extracting correlation laws for these configurations.

3.6. Acknowledgments

This work was granted access to the HPC resources of CINES under the allocation A0032b06115 made by GENCI (Grand Equipement National de Calcul Intensif) and to the resources of CALMIP supercomputing center under the allocation 2017-P1529.

3.7. References

Adanez, J., Abad, A., Garcia-Labiano, F., Gayan, P. and de Diego, L.F. (2012). Progress in chemical-looping combustion and reforming technologies [Online]. *Progress in Energy and Combustion Science*, 38(2), 215–282. Available at: http://linkinghub.elsevier.com/retrieve/pii/S0360128511000505.

Ali, H. and Rohani, S. (1997). Dynamic modeling and simulation of a riser-type fluid catalytic cracking unit. *Chemical Engineering and Science*, 20, 118–130.

Aslam, T.D. (2004). A partial differential equation approach to multidimensional extrapolation [Online]. *Journal of Computational Physics*, 193(1), 349–355. Available at: http://linkinghub.elsevier.com/retrieve/pii/S0021999103004170.

Bagheri, G., Bonadonna, C., Manzella, I. and Vonlanthen, P. (2015). On the characterization of size and shape of irregular particles [Online]. *Powder Technology*, 270, 141–153. Available at: http://linkinghub.elsevier.com/retrieve/pii/S0032591014008651.

Beetstra, R., van der Hoef, M.A. and Kuipers, J.A.M. (2007). Drag force of intermediate Reynolds number flow past mono- and bidisperse arrays of spheres [Online]. *AIChE Journal*, 53(2), 489–501. Available at: http://doi.wiley.com/10.1002/aic.11065.

Caltagirone, J.-P. and Vincent, S. (2001). Tensorial penalisation method for solving Navier–Stokes equations. *Comptes Rendus de l'Académie des Sciences Paris – Série IIb*, 329, 607–613.

Chadil, M.-A., Vincent, S. and Estivalèzes, J.-L. (2018a). Accurate calculation of heat transfer coefficients for motions around particles with a finite-size particle approach. *Turbulence and Interactions. Proceedings of the TI 2018 Conference*, June 25–29, Les Trois-Îlets, Martinique, France.

Chadil, M.-A., Vincent, S. and Estivalèzes, J.-L. (2018b). Drag, lift and Nusselt coefficients for ellipsoid using Particle-Resolved Direct Numerical Simulations. *Turbulence and Interactions. Proceedings of the TI 2018 Conference*, June 25–29, Les Trois-Îlets, Martinique, France.

Chadil, M.-A., Vincent, S. and Estivalèzes, J.-L. (2018c). Accurate estimate of drag forces using particle-resolved direct numerical simulations [Online]. *Acta Mechanica*, 230(2), 569–595. Available at: http://link.springer.com/10.1007/s00707-018-2305-1.

Chadil, M.-A., Vincent, S. and Estivalèzes, J.-L. (2019). Improvement of the viscous penalty method for particle-resolved simulations [Online]. *Open Journal of Fluid Dynamics*, 9(2), 168–192. Available at: https://hal.archives-ouvertes.fr/hal-02641430.

Clair, B.P.L., Hamielec, A.E. and Pruppacher, H.R. (1970). A numerical study of the drag on a sphere at low and intermediate Reynolds numbers [Online]. *Journal of the Atmospheric Sciences*, 27(2), 308–315. Available at: https://doi.org/10.1175/1520-0469(1970)027<0308:ANSOTD>2.0.CO;2.

Dartevelle, S. (2007). From model conception to verification and validation, a global approach to multiphase Navier–Stoke models with an emphasis on volcanic explosive phenomenology [Online]. Technical Report. Available at: http://www.osti.gov/servlets/purl/948564-J3qrTG/.

Deen, N.G., Kriebitzsch, S.H., van der Hoef, M.A. and Kuipers, J. (2012). Direct numerical simulation of flow and heat transfer in dense fluid-particle systems [Online]. *Chemical Engineering Science*, 81, 329–344. Available at: http://linkinghub.elsevier.com/retrieve/pii/S0009250912004204.

Deen, N.G., Peters, E.A.J.F., Padding, J.T. and Kuipers, J.A.M. (2014). Review of direct numerical simulation of fluid-particle mass, momentum and heat transfer in dense gas-solid flows [Online]. *Chemical Engineering Science*, 116, 710–724. Available at: http://linkinghub.elsevier.com/retrieve/pii/S0009250914002656.

Dennis, S.C.R. and Walker, J.D.A. (1971). Calculation of the steady flow past a sphere at low and moderate Reynolds numbers. *Journal of Fluid Mechanics*, 48(4), 771–789.

Ergun, S. (1952). Fluid flow through packed columns. *Chemical Engineering Progress*, 48, 89–94.

Feng, Z.-G. and Michaelides, E.E. (2000). A numerical study on the transient heat transfer from a sphere at high Reynolds and Peclet numbers. *International Journal of Heat and Mass Transfer*, 43(2), 219–229.

Fortin, M. and Glowinski, R. (1982). *Méthodes de lagrangien augmenté. Application à la résolution numérique de problèmes aux limites*. Dunod, Paris.

Gobin, A., Neau, H., Simonin, O., Llinas, J.-R., Reiling, V. and Sélo, J.-L. (2003). Fluid dynamic numerical simulation of a gas phase polymerization reactor. *International Journal for Numerical Methods in Fluids*, 43(10–11), 1199–1220.

Gunn, D. (1978). Transfer of heat or mass to particles in fixed and fluidized beds [Online]. *International Journal of Heat and Mass Transfer*, 21(4), 467–476. Available at: http://www.sciencedirect.com/science/article/pii/0017931078900807.

Hill, R.J., Koch, D.L. and Ladd, A.J.C. (2001). Moderate-Reynolds-number flows in ordered and random arrays of spheres [Online]. *Journal of Fluid Mechanics*, 448. Available at: http://www.journals.cambridge.org/abstract_S0022112001005936.

Hoef, M.A.V.D., Beetstra, R. and Kuipers, J.A.M. (2005). Lattice-Boltzmann simulations of low-Reynolds-number flow past mono- and bidisperse arrays of spheres: Results for the permeability and drag force [Online]. *Journal of Fluid Mechanics*, 528. Available at: http://www.journals.cambridge.org/abstract_S0022112004003295.

Kataoka, I. (1986). Local instant formulation of two-phase flow. *International Journal of Multiphase Flow*, 12, 745–758.

Ladd, A.J.C. (1994a). Numerical simulations of particulate suspensions via a discretized Boltzmann equation: Part 1. Theoretical foundation [Online]. *Journal of Fluid Mechanics*, 271, 285. Available at: http://www.journals.cambridge.org/abstract_S0022112094001771.

Ladd, A.J.C. (1994b). Numerical simulations of particulate suspensions via a discretized Boltzmann equation. Part 2: Numerical results [Online]. *Journal of Fluid Mechanics*, 271, 311. Available at: http://www.journals.cambridge.org/abstract_S0022112094001783.

Lyngfelt, A., Leckner, B. and Mattisson, T. (2001). A fluidized-bed combustion process with inherent CO2 separation – Application of chemical-looping combustion. *Chemical Engineering Science*, 56(10), 3101–3113.

Magnaudet, J., Rivero, M. and Fabre, J. (1995). Accelerated flows past a rigid sphere or a spherical bubble. Part 1. Steady straining flow [Online]. *Journal of Fluid Mechanics*, 284, 97. Available at: http://www.journals.cambridge.org/abstract_S0022112095000280.

Massol, A. (2004). Simulations numériques d'écoulements à travers des réseaux fixes de sphères monodisperses et bidisperses, pour des nombres de Reynolds modérés. PhD Thesis, Toulouse Institute of Technology, France.

Mattisson, T., Keller, M., Linderholm, C., Moldenhauer, P., Rydan, M., Leion, H. and Lyngfelt, A. (2018). Chemical-looping technologies using circulating fluidized bed systems: Status of development [Online]. *Fuel Processing Technology*, 172, 1–12. Available at: http://linkinghub.elsevier.com/retrieve/pii/S0378382017311311.

de Motta, J.C.B., Breugem, W., Gazanion, B., Estivalèzes, J.-L., Vincent, S. and Climent, E. (2013). Numerical modelling of finite-size particle collisions in a viscous fluid. *Physics of Fluids*, 25, 083302.

Ozel, A., de Motta, J.C.B., Abbas, M., Fede, P., Masbernat, O., Vincent, S., Estivalèzes, J.-L. and Simonin, O. (2017). Particle resolved direct numerical simulation of a liquid – Solid fluidized bed: Comparison with experimental data. *International Journal of Multiphase Flow*, 89, 228–240.

Randrianarivelo, T., Pianet, G., Vincent, S. and Caltagirone, J.-P. (2005). Numerical modelling of the solid particle motion using a new penalty method. *International Journal for Numerical Methods in Fluids*, 47, 1245–1251.

Ranz, W.E. and Marshall, W.R. (1952). Evaporation from drops. Parts I & II. *Chemical Engineering Progress*, 48, 141–146; 173–180.

Sarthou, A., Vincent, S., Angot, P., Caltagirone, J.-P. (2009). The sub-mesh penalty method [Online]. *Finite Volumes for Complex Applications V*. Available at: http://hal.archives-ouvertes.fr/hal-00390080/.

Schiller, L. and Naumann, A.Z. (1933). Über die grundlegenden Berechnungen bei der Schwerkraftaufbereitung. *Zeitschrift des Vereines Deutscher Ingenieure*, 77, 318–320.

Schönfeld, T. and Rudgyard, M. (1999). Steady and unsteady flow simulations using the hybrid flow solver AVBP [Online]. *AIAA Journal*, 37(11), 1378–1385. Available at: https://doi.org/10.2514/2.636.

Sun, B., Tenneti, S. and Subramaniam, S. (2015). Modeling average gas–solid heat transfer using particle-resolved direct numerical simulation [Online]. *International Journal of Heat and Mass Transfer*, 86, 898–913. Available at: http://www.sciencedirect.com/science/article/pii/S001793101500304X.

Sweby, P.K. (1984). High resolution schemes using flux limiters for hyperbolic conservation laws [Online]. *SIAM Journal on Numerical Analysis*, 21(5), 995–1011. Available at: http://www.jstor.org/stable/2156939.

Tavassoli, H., Kriebitzsch, S., van der Hoef, M., Peters, E. and Kuipers, J. (2013). Direct numerical simulation of particulate flow with heat transfer [Online]. *International Journal of Multiphase Flow*, 57, 29–37. Available at: http://linkinghub.elsevier.com/retrieve/pii/S0301932213001055.

Tenneti, S., Garg, R. and Subramaniam, S. (2011). Drag law for monodisperse gas–solid systems using particle-resolved direct numerical simulation of flow past fixed assemblies of spheres [Online]. *International Journal of Multiphase Flow*, 37(9), 1072–1092. Available at: http://linkinghub.elsevier.com/retrieve/pii/S0301932211001170.

Tenneti, S., Sun, B., Garg, R. and Subramaniam, S. (2013). Role of fluid heating in dense gas–solid flow as revealed by particle-resolved direct numerical simulation [Online]. *International Journal of Heat and Mass Transfer*, 58(1–2), 471–479. Available at: http://www.sciencedirect.com/science/article/pii/S0017931012008526.

Uhlmann, M. (2005). An immersed boundary method with direct forcing for the simulation of particulate flows. *Journal of Computational Physics*, 209(2), 448–476.

Uhlmann, M. (2008). Interface-resolved direct numerical simulation of vertical particulate channel flow in the turbulent regime. *Physics of Fluids*, 20(5), 053305.

Vincent, S., Caltagirone, J.-P., Lubin, P. and Randrianarivelo, N. (2004). An adaptative augmented Lagrangian method for three-dimensional multi-material flows. *Computers and Fluids*, 33, 1273–1289.

Vincent, S., Randrianarivelo, T., Pianet, G. and Caltagirone, J. (2007). Local penalty methods for flows interacting with moving solids at high Reynolds numbers. *Computers and Fluids*, 36, 902–913.

Vincent, S., Sarthou, A., Caltagirone, J.-P., Sonilhac, F., Février, P., Mignot, C. and Pianet, G. (2011). Augmented Lagrangian and penalty methods for the simulation of two-phase flows interacting with moving solids. Application to hydroplaning flows interacting with real tire tread patterns. *Journal of Computational Physics*, 230, 956–983.

Vincent, S., de Motta, J.C.B., Sarthou, A., Estivalèzes, J.-L., Simonin, O. and Climent, E. (2014). A Lagrangian VOF tensorial penalty method for the DNS of resolved particle-laden flows. *Journal of Computational Physics*, 256, 582–614.

Wang, Z., Fan, J. and Luo, K. (2008). Combined multi-direct forcing and immersed boundary method for simulating flows with moving particles [Online]. *International Journal of Multiphase Flow*, 34(3), 283–302. Available at: http://www.sciencedirect.com/science/article/pii/S0301932207001474.

Wen, C.-Y. and Yu, Y.H. (1966). Mechanics of fluidization. *Chemical Engineering Progress Symposium Series*, 62, 100–111.

Whitaker, S. (1972). Forced convection heat transfer correlations for flow in pipes, past flat plates, single cylinders, single spheres, and for flow in packed beds and tube bundles. *AIChE Journal*, 18(2), 361–371.

Zastawny, M., Mallouppas, G., Zhao, F. and van Wachem, B. (2012). Derivation of drag and lift force and torque coefficients for non-spherical particles in flows. *International Journal of Multiphase Flow*, 39, 227–239.

4

Interfaces and Critical Fluids

For pure substances, the liquid–vapor phase separation interface disappears at the critical point. This property, among others, has attracted the interest of many scientists. This chapter is an updated version of the text of the presentation on critical fluids presented at a seminar at Saint-Gobain-Recherche Aubervilliers on June 14, 2007. First, we present thermostatics; then, in order, the thermodynamics of fluids near the critical point; the piston effect, a specific mode of heat transfer; the expansion of a "drop" at the critical pressure; the behavior of a supercritical fluid pocket immersed in a high-temperature environment; and, finally, boiling near the critical point. It partly revisits texts published by the various authors cited.

List of symbols

Symbol	Meaning	Symbol	Meaning
a, b	Van der Waals constants	α, β, γ, δ, ν	Exponents of the renormalization group
c_p	Specific heat at constant pressure	λ	Thermal conductivity
D	Diameter of the drop	$\kappa=\lambda/\rho c_p$	Thermal diffusivity
$F, \overline{F}$	Free energy, molar Helmholtz free energy	μ	Shear viscosity
$K_T = (\partial \ln \rho / \partial P)_T$	Isothermal compressibility	p, p_c	Density, critical density

Chapter written by Roger PRUD'HOMME.

Symbol	Meaning	Symbol	Meaning
p, p_c	Pressure, critical pressure	σ	Surface tension
R	Molar constant of perfect gases	ξ	Correlation length
S	Entropy	τ	Dimensionless time
T, T_c	Temperature in K, critical temperature	$v, \bar{v}$	Volume, molar volume

It is generally quite easy to distinguish between the different phases of a pure body. However, there may be some situations where there is greater difficulty. For example, for a pure fluid subject to sufficient pressure, the usual meniscus that separates the liquid and gaseous phases may disappear, giving way to a veritable continuity between these two phases. This is known as the supercritical fluid state (Zappoli *et al.* 2015) (Figure 4.1).

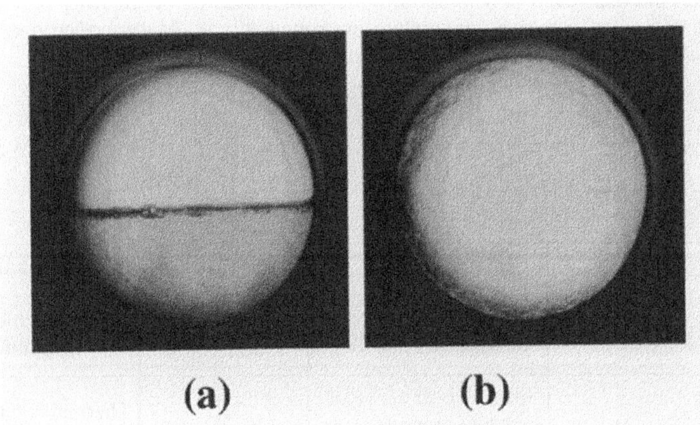

Figure 4.1. *(a) The two phases (liquid and gas) of CO2 are clearly seen (meniscus); b) with an increase in temperature, when the critical temperature and pressure are reached, it is no longer possible to differentiate between the two phases (liquid and gas) and there is only one single phase: the supercritical fluid (Department of Chemistry, Leeds University). For a color version of this figure, see www.iste.co.uk/prudhomme/fluid1.zip*

The topic of critical and supercritical fluids (SCF) is of fundamental interest, and the potential fields of application of supercritical fluids range from the field of space launch vehicles to the field of organic waste processing (Beysens 2001).

Supercritical fluids, which are very compressible and expandable near the critical point, are also very dense and sensitive to gravity, so much so that *experiments must often be carried out in microgravity to understand their behavior and model them*. On the other hand, at the critical point, certain physical properties diverge (such as isothermal compressibility, $K_T = (\partial \ln \rho / \partial P)_T$, specific heat c_p at constant pressure and thermal conductivity λ) or tend towards zero (e.g. thermal diffusivity $k = \lambda / \rho c_p$, heat transmission through conductivity slows down considerably, known as "**critical slowing**") and new phenomena appear. Experiments carried out in microgravity have made it possible to confirm the existence of the "**piston effect**", which is a new mode of heat transmission that is demonstrated through theory (Zappoli *et al.* 1990).

4.1. Thermostatics of fluids in the vicinity of the critical point

The elements given below are related to the laws of state of real fluids. Perfect gases, i.e. gases under low pressure, are well known. On the other hand, the behaviors of gases under high pressures are less well known. Nonetheless, we will see that it is possible to use the behavior of the perfect gas of a fluid to arrive at the form its laws of state will take at higher pressures.

4.1.1. *Real fluids*

The state diagram of a pure body is schematically represented in Figure 4.2. The domain of perfect gases corresponds to low pressures and high volumes. At the critical liquid vapor point C, the first and second partial derivatives of pressure with respect to volume at constant temperature cancel each other out:

$$\left(\partial p / \partial \bar{V}\right)_T = 0, \quad \left(\partial^2 p / \partial \bar{V}^2\right)_T = 0 \qquad [4.1]$$

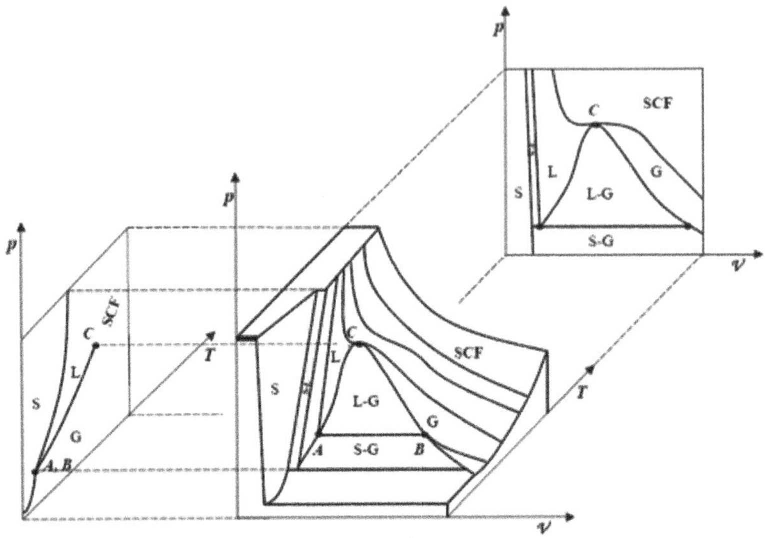

Figure 4.2. *State diagram of a pure body, S: solid state, L: liquid state, G: gaseous state, SCF: supercritical fluid*

Table 4.1 gives the coordinates of the critical point of some bodies (Garrabos *et al.* 1992).

Body	Critical temperature T_c (K)	Critical pressure p_c (MPa)	Critical density (kg/m^3)
Water	647.1	22.06	322.2
Carbon dioxide	304.14	7.378	467.8
Oxygen	154.58	5.043	436.2
Nitrogen	126.24	3.398	313.9
Ammonia	405.4	11.1	235
Helium	3.316	0.114	41.45

Table 4.1. *Coordinates of the critical point of some bodies (Garrabos* et al. *1992)*

The law of state being generally given in the form $p = p(T, \overline{\mathcal{V}})$, where T is the temperature and $\overline{\mathcal{V}}$ is the molar volume, we will define a standard molar volume $\overline{\mathcal{V}}^0$ corresponding to a standard temperature state *T* where the gas is perfect and where the standard pressure p^0 is such that $p^0\ \overline{\mathcal{V}}^0 = RT^0$. Such a standard state always exists.

NOTE.– It is possible to use the following method to determine the form of the thermodynamic functions of a real gas, provided that these functions are known for the perfect gas.

Starting from the equation of state, it is possible to determine the Helmholtz free energy as a function of $T, \overline{\mathcal{V}}$ and the other thermodynamic functions, using an **approximation method**. In order to do this, we must accept the existence of a standard reference state defined by the molar volume $\overline{\mathcal{V}}^0$ such that, for $\overline{\mathcal{V}} > \overline{\mathcal{V}}^0$, the fluid may be considered as a perfect gas, i.e. $p\overline{\mathcal{V}} = RT$. Since $\partial\overline{F}/\partial\overline{\mathcal{V}} = -\,p(T, \overline{\mathcal{V}})$, it is possible to write $\overline{F} = \breve{F}_T^0 - \int_{\overline{\mathcal{V}}^0}^{\overline{\mathcal{V}}} p\, d\mathcal{V}$. This gives the standard value of molar free energy $\breve{F}_T^0$ of the perfect gas at $\overline{\mathcal{V}} = \overline{\mathcal{V}}^0$. We write $\int_{\overline{\mathcal{V}}^0}^{\overline{\mathcal{V}}} p\, d\mathcal{V} = \int_{\overline{\mathcal{V}}^0}^{\overline{\mathcal{V}}} \left(p - RT/\overline{\mathcal{V}}\right) d\mathcal{V}$ $+ RT \ln\left(\overline{\mathcal{V}}/\overline{\mathcal{V}}^0\right)$. The integral may be written as $\int_{\overline{\mathcal{V}}^0}^{\overline{\mathcal{V}}} \left(p - RT/\overline{\mathcal{V}}\right) d\mathcal{V}$ $= \int_{\overline{\mathcal{V}}^0}^{\infty} \left(p - RT/\overline{\mathcal{V}}\right) d\mathcal{V} + \int_{\infty}^{\overline{\mathcal{V}}} \left(p - RT/\overline{\mathcal{V}}\right) d\mathcal{V} \cong \int_{\infty}^{\overline{\mathcal{V}}} \left(p - RT/\overline{\mathcal{V}}\right) d\mathcal{V}$, since, for $\overline{\mathcal{V}} > \overline{\mathcal{V}}^0$, the fluid is considered to be a perfect gas. From this, we derive $\overline{F} \cong \breve{F}_T^0 - RT \ln\left(\overline{\mathcal{V}}/\overline{\mathcal{V}}^0\right) - \int_{\infty}^{\overline{\mathcal{V}}} \left(p - RT/\overline{\mathcal{V}}\right) d\mathcal{V}$.

4.1.2. *A van der Waals fluid*

The van der Waals equation of state for a fluid is

$$\left(\overline{\mathcal{V}} - p\right)\left(p + a/\overline{\mathcal{V}}^2\right) = RT \qquad [4.2]$$

This equation of state is simplest for real gases and liquids. It takes into account a minimum distance between the molecules and interaction between these molecules. For rarefied gases: $a/\overline{\mathcal{V}}^2 << p$, hence: $\overline{\mathcal{V}} - b \cong \left(RT/p\right)$ $\left(1 - a/p\overline{\mathcal{V}}^2\right)$, and **for perfect gases** (limiting case $\overline{\mathcal{V}} >> b - aRT/p^2\,\overline{\mathcal{V}}^2$):

$\bar{\mathcal{V}} = RT/p$. For liquids: $a/\bar{\mathcal{V}}^2 >> p$ ($a/\bar{\mathcal{V}}^2$ some tens of thousands of atmospheres), it is often possible to neglect p, such that $\bar{\mathcal{V}} \cong \bar{\mathcal{V}}_T^0$.

NOTE.– With the reduced variables $x = \bar{\mathcal{V}}/\bar{\mathcal{V}}_c$, $y=p/p_c$, $z=T/T_c$, the van der Waals equation of state can be written as $y = \dfrac{8\,z}{3x-1} - \dfrac{3}{x^2}$. Using these variables, the perfect gas equation becomes $y_{pft} = 8z/3x$.

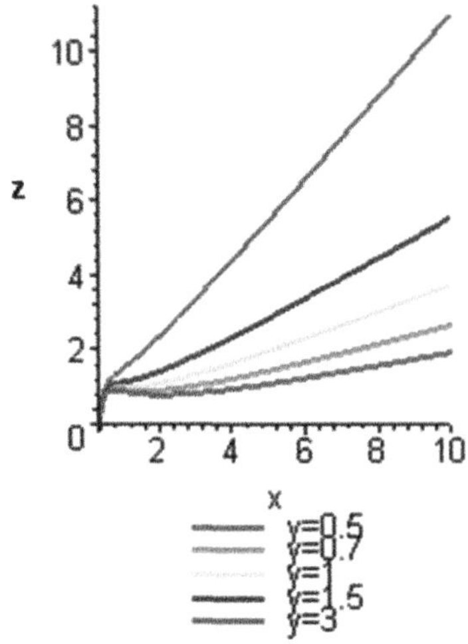

Figure 4.3. *Definition of the standard state: view of isobars z(x) for y = p/p_c = const. For a color version of this figure, see www.iste.co.uk/prudhomme/fluid1.zip*

The fluid may be considered a perfect gas if $y - y_{pft} << 1$. The isobars $z = (3x - 1)(y + 3 / x^2)/8$, y = const. are represented in Figure 4.3. It can be seen that these curves tend to become straight lines when the relative volume x increases. This proves that, for a van der Waals fluid, over any given isothermal temperature T, it is possible to define a standard state, provided that the values of the parameters are chosen in the right range. This result, which is related to the definition of a perfect gas, does not depend on the fluid being considered.

Let us now directly determine the forms of internal energy $\overline{E}$, entropy $\overline{S}$ and free molar enthalpy of a van der Waals fluid. It is possible to write, successively (Prud'homme 1988):

$$p=\frac{RT}{\overline{\mathcal{V}}-b}-\frac{a}{\overline{\mathcal{V}}^2}, d\overline{S}=\frac{1}{T}d\overline{E}+\frac{p}{T}d\overline{\mathcal{V}}+\frac{R}{\overline{\mathcal{V}}-b}d\overline{\mathcal{V}}-\frac{a}{T\overline{\mathcal{V}}^2}d\overline{\mathcal{V}}$$

$$=\frac{1}{T}d\left(\overline{E}\frac{a}{\frac{1}{T}dE}\right)+Rd\ \ln\left(\overline{\mathcal{V}}-b\right)$$

$$\overline{S}=\overline{S}\left(\overline{E}+\frac{a}{\overline{\mathcal{V}}},\overline{\mathcal{V}}-b\right), \text{hence } \frac{\partial R}{\partial\left(\overline{E}+a/\overline{\mathcal{V}}\right)}=0=\frac{\partial(1/T)}{\partial\ln\left(\overline{\mathcal{V}}-b\right)}\mathcal{V}$$

It follows that, with $\overline{\mathcal{V}}^o$ corresponding to a standard reference state, since T is not related to $\overline{\mathcal{V}}-b$

$$\overline{E}=\overline{E}_T^0-a\ \left(1/\overline{\mathcal{V}}-1/\overline{\mathcal{V}}^0\right) \qquad [4.3]$$

We have

$\overline{S}=\overline{S}_T^0+R\ln\left[\overline{\mathcal{V}}-b/\left(\overline{\mathcal{V}}^o-b\right)\right]$ with $\overline{S}_T^0=\overline{S}_0^0+\int_{T^0}^T d\overline{E}_T^0/T$, since $d\overline{S}=d\left(\overline{E}_T^0/T\right)+Rd\ \ln\left(\overline{\mathcal{V}}-b\right)$

Finally, the expression for $\overline{F}$ is obtained, from which the expression for molar free enthalpy μ can be derived:

$$\overline{F}=\breve{F}_T^0-a\left(1/\overline{\mathcal{V}}-1/\overline{\mathcal{V}}^o\right)-RT\ \ln\ \left(\left(\overline{\mathcal{V}}\text{-}b\right)/\overline{\mathcal{V}}^o\text{-}b\right),\ \breve{F}_T^0=\overline{E}_T^0-T\ \breve{S}_T^0, \qquad [4.4]$$

$$\mu=\overline{E}-T\overline{S}+p\overline{\mathcal{V}}=\breve{\mu}_T^0-\ 2a\left(\ 1/\overline{\mathcal{V}}-\ 1/\overline{\mathcal{V}}^0\right)-RT\ \ln\ \left(\left(\overline{\mathcal{V}}\text{-}b\right)/\left(\overline{\mathcal{V}}^o\text{-}b\right)\right)+ RT\ \left(\overline{\mathcal{V}}/\left(\overline{\mathcal{V}}\text{-}b\right)-\overline{\mathcal{V}}^o/\left(\overline{\mathcal{V}}^o\text{-}b\right)\right)$$

NOTE.– If the approximation method from section 4.1.1 is applied, the formula $\bar{F} \cong \breve{F}_T^0 - RT \ln\left(\bar{\mathcal{V}}/\bar{\mathcal{V}}^{\circ}\right) - RT \ln\left(\left(\bar{\mathcal{V}} - b\right)/\bar{\mathcal{V}}\right) - a\left(1/\bar{\mathcal{V}}\right)$ is found for Helmholtz free energy. This expression is slightly different from equation [4.4].

4.1.3. *Other laws for gases and dense liquids*

The following form is applicable to the van der Waals, Redlich–Kwong and Peng–Robinson laws for the equations of state (Prausnitz *et al.* 1999; Reid *et al.* 1987):

$$p = \frac{RT}{\bar{\mathcal{V}} - b} \frac{a}{\bar{\mathcal{V}}^2 + ub\bar{\mathcal{V}} + wb^2} \qquad [4.5]$$

where u and w are coefficients.

NOTE.– In the case of the Peng–Robinson equation

$$u=2,\ w=-1,\ b=\frac{0.07780RT_c}{p_c},\ a=\frac{0.45724R^2T_c^2}{p_c}\left[1+f\omega\left(1-\left(\frac{T}{T_c}\right)^{½}\right)\right]^2,$$

$$f\omega=0.37464+1.54226\omega-0.26992\omega^2$$

where the acentric factor ω is tabulated for many species. Thus: $\dfrac{\partial^2 p}{\partial T^2} = -\dfrac{d^2a/dT^2}{\bar{V}^2 + 2b\bar{V} - b^2}$.

Reid, Prausnitz and Poling provide the data necessary to obtain the equation of state $p = p\left(T, \bar{\mathcal{V}}\right)$ of some bodies and also to determine the molar thermodynamic functions in the ideal gas state (calorific capacities $\bar{C}_{p,j}$ in J/mol.K, the heat of standard formations $\left(\bar{Q}_f^0\right)_j$ at 298.2 K in J/mole and free enthalpies of formation $\left(\mu_0^0\right)_j$ at 289.2 K under 1 bar in J/mole).

4.2. Thermodynamics of fluids in the vicinity of the critical point

4.2.1. *Universal exponents*

In thermodynamics, reversible and irreversible evolutions are considered. In order to study a thermodynamic evolution, we need both thermostatic laws and additional laws that bring in transfer coefficients. In general, these laws are known for conditions that are far from the critical point conditions.

The closer we get to the critical point, the less valid the conventional laws are.

These conventional laws (for instance, the van der Waals laws) arise from the **mean field theory**. The term “mean field” comes from the property that is common to these models, assuming that molecular interactions only act on a given molecule, on average, without involving the position of the molecule with respect to its neighbors. This is a statistical physics hypothesis that does not take into account the spatial distribution of molecules in the analysis of molecular interactions. Using this theory, it is possible to obtain an easy-to-use analytical expression, but these have not been well verified in the vicinity of the critical point.

Thus, the conditions at the critical point $\left(\frac{\partial p}{\partial \overline{\mathcal{V}}}\right)_T = \left(\frac{\partial^2 p}{\partial \overline{\mathcal{V}}^2}\right)_T = 0$ lead to the relations:

$\overline{\mathcal{V}}_c = 3b$, $T_c = 8a/27Rb$, $p_c = a/27b^2$, from which we derive: $a = 9/8RT_c\overline{\mathcal{V}}_c$, $b = \overline{\mathcal{V}}_c/3$ and also $p_c = \frac{3}{8}\frac{RT_c}{\overline{V}_v}$. This last relation has not been verified under real conditions (see Table 4.2).

In the case of a van der Waals fluid, the isothermal compressibility K_T becomes:

$$K_T = -\frac{4\overline{\mathcal{V}}_c^2}{9RT_c}\left(\frac{T - T_c}{T_c}\right)^{-1}, \qquad [4.6]$$

a law which has not been verified under real conditions, as demonstrated in the first row of Table 4.2.

Properties	**Thermodynamic pathway**	**Expression**	**Exponent of the renormalization group theory**	**Exponent of the mean field theory**
Isothermal compressibility	$\rho = \rho_c$	$K_T = \Gamma \left\lvert \frac{T - T_c}{T_c} \right\rvert^{-\gamma}$	γ=1.239±0.002	$\gamma = 1$
Specific heat at constant volume	$\rho = \rho_c$	$C_v = A \left\lvert \frac{T - T_c}{T_c} \right\rvert^{-\alpha}$	α=0.110±0.003	$\alpha = 0$
Liquid/ gas-particle density jump	Coexistence curve	$\rho_l - \rho_g = 2B \left\lvert \frac{T - T_c}{T_c} \right\rvert^{\beta}$	β=0.325±0.002	$\beta = 0.5$
Correlation length	$\rho = \rho_c$	$\xi = \xi_0 \left\lvert \frac{T - T_c}{T_c} \right\rvert^{-\nu}$	ν=0.630±0.001	$\nu = 0.5$
Pressure	$T = T_c$	$p - p_c \propto (\frac{p - p_c}{p_c})\delta$	δ=4.8±0.02	$\delta = 3$

Table 4.2. *Some properties in the vicinity of a critical point and the associated critical exponents (Sengers 1978). Γ, A, B and ξ_0 are the cofactors that characterize the substance studied*

On the other hand, the **renormalization group theory** makes it possible to derive, in particular, the behavior of transfer coefficients near the critical point. It also gives the behaviors of other thermodynamic properties.

Let ϕ be a physical quantity. We consider an evolution along a given thermodynamic pathway (e.g. with the constant density equal to critical density ρ_c). The physical quantity ϕ is then a function only of the relative temperature T / T_c. Formulas in the following form are obtained:

$$\phi / \phi_r \propto (T / T_c - 1)^s \qquad [4.7]$$

where s is a constant, universal coefficient, valid for a given type of critical transformation and a given parameter, ϕ_r, a reference value for ϕ.

– If s is positive, the considered quantity tends towards zero when the temperature tends towards its critical value. *This is the case for surface tension*, which is canceled out at the critical point (Mezger 1946; Bachaud 1951).

This is also the case for the compressibility at constant temperature $K_T = (\partial \ln \rho / \partial p)_T$, which can be easily derived from the van der Waals equation of state, as has been seen.

– If s is negative, there is divergence at the critical point. This is the case for thermal conductivity λ, which, thus, tends towards infinity. The calorific capacity at constant pressure c_p is also a divergent quantity.

However, thermal diffusivity $K=\lambda / \rho c_p$ is canceled out as it approaches the critical point. According to this result, heat transmission through conductivity becomes almost impossible. Another mode of heat transmission comes into play, called the "piston effect".

4.2.2. *Isobaric evolutions*

At $p = p_c$, we have, according to Arias-Zugasti (Arias-Zugasti *et al.* 1999):

$$\rho - \rho_c \cong -\rho_0 \left(\frac{T - T_c}{T_c} \right)^{1/\delta}, c_p \cong c_0 \left| \frac{T - T_c}{T_c} \right|^{-\gamma/\beta\delta}, \lambda \cong \rho_c \kappa_0 c_0 \left| \frac{T - T_c}{T_c} \right|^{-(\gamma-\nu)/\beta\delta} \quad [4.8]$$

4.2.3. *Valid expressions of varying distance from the critical point*

Consider an isobaric evolution (Arias-Zugasti *et al.* 1999). The following is a valid expression for thermal conductivity of varying distance from the critical point:

$$\lambda = \lambda^{gas}(T) + \lambda^{E}(\rho) + \lambda^{C}(T) \quad [4.9]$$

where $\lambda^{gas}(T)$ is the limit at low density, $\lambda^{E}(\rho)$ is the excessive thermal conductivity and $\lambda^{C}(T)$ is the critical divergence.

– $\lambda^{gas}(T) \equiv \lambda_\infty T / T_\infty$ (a linear function of temperature), or $\lambda^{gas}(T) \equiv \lambda_\infty \sqrt{T / T_\infty}$ (given by the kinetic theory of gases).

– $\lambda^{E}(\rho) = \lambda_1 \rho / \rho_c$ (linear dependence on density).

– $\lambda^C(T) = \lambda_0 \left| \frac{T - T_c}{T_c} \right|^{-1/3}$ (mean field theory).

4.2.4. *Summary of theoretical approaches*

Two different theoretical approaches are commonly used to understand the singular behavior of transport coefficients at the critical point (Luettmer-Strathmann *et al.* 1995):

– the **dynamic renormalization group theory** mentioned above, first formulated by Halpering *et al.*, is an extension of the ideas in Wilson and Fisher's static renormalization group theory of dynamic critical phenomena. Its strengths lie in its ability to predict asymptotic critical behavior of the transport coefficients and the elucidation of phenomenological concepts, such as dynamic scales and dynamic universality;

– the **mode coupling theory** came from Fixman's hypothesis that critical anomalies of transport coefficients could be understood as resulting from nonlinear couplings between the hydrodynamic modes of the system. One advantage of the mode coupling theory is that it can easily be applied outside the near-critical region. Starting from the work of Olchowy and Gelbart, the mode coupling calculations and Ferrel and Perl's relative theory of decoupled modes were used to yield expressions for transport coefficients that were valid outside the critical region (Sengers and Levelt Sengers 1978; Sengers 1985).

4.3. A specific mode of heat transmission: the piston effect

Thermal diffusivity $\kappa = \lambda / \rho c_p$ tends towards zero at the critical liquid vapor point. Thus it is expected that the heat transport will be very slow. This is not the case, however, as is demonstrated in the results obtained in experiments that were carried out in space (Nitsche and Straub 1987).

Indeed, experiments in microgravity are necessary for understanding phenomena in critical fluids, as the high dilatability of these fluids is the reason for the large variations in densities, which produce intense convective motions that hinder precise observations.

In 1990, three different independent teams, Zappoli *et al.* (1990), Boukari *et al.* (1990) and Onuki *et al.* (1990) explained this phenomenon. Onuki noted that the heat transfer equation $\frac{\partial T}{\partial t} - \kappa \nabla^2 T = 0$ was no longer relevant for a fluid maintained at a fixed volume. A change in entropy at any point in the volume of the fluid leads to a change in pressure, which, in turn, produces an adiabatic change in temperature throughout the volume. He emphasized that this phenomenon had already been described in the literature, but, as it was negligible for a fluid at normal compressibility, it was also unfortunately neglected in the study of fluids close to their critical points.

In an incompressible fluid, the variations in internal energy are due only to temperature variations that result from heat flows imposed on the fluid, while, for a compressible fluid, adiabatic variations of temperature and density, caused by variations in pressure over time, must be taken into account (Boukari *et al.* 1990). Adiabatic effects are dominant in a fluid close to its critical point, due to the divergence of its compressibility.

A new heat transfer mechanism was demonstrated within supercritical fluids: the piston effect, which acts when a cell that is filled with critical fluid is subjected to local heating.

The French team (Zappoli *et al.* 1990; Prud'homme 2016) chiefly used the singular perturbation methods, applied to Navier–Stokes equations for a van der Waals fluid.

The piston effect mechanism works as follows: when a wall is heated, the heat diffuses from its source over a narrow area, and the fluid subjected to this heating expands strongly and isentropically compresses the rest of the fluid, which then undergoes a homogeneous rise in temperature. The expanded fluid contained in the boundary layer acts as a piston, leading to the creation of a system of compression waves, which are propagated throughout the fluid, and the first response of the system to a thermal perturbation over a narrow area is, therefore, mechanical. The repetitive passage of the continually produced acoustic field causes the appearance of a low velocity of the fluid volume as well as total compression of the entire volume. The piston effect is, therefore, the result of a thermomechanical coupling, as depicted in Figure 4.4:

a) the supercritical fluid contained in a closed cell is heated at one end, the heat brought to the end diffuses into a thin thermal boundary layer, in which very strong temperature gradients appear (Figure 4.4a);

b) the high compressibility of the fluid, causing the expansion of the hot layer, compresses the rest of the fluid in an adiabatic manner (Figure 4.4b);

c) the expansion of the hot layer acts like a piston on the rest of the fluid, hence the name "the piston effect". A rapid and homogeneous increase in temperature throughout the cell is observed (Figure 4.4c).

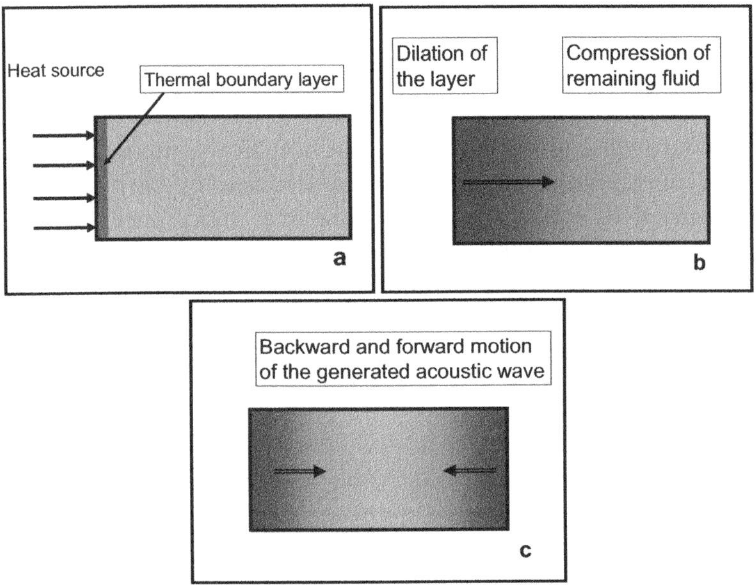

Figure 4.4. *Schematic representation of the piston effect mechanism (the dark level symbolizes the rise in temperature). For a color version of this figure, see www.iste.co.uk/prudhomme/fluid1.zip*

In summary, through a purely dynamic mechanism, a fluid that diffuses heat very poorly can nevertheless transport thermal energy at high speed. The temperature relaxation is therefore increasingly rapid in a perfectly isobaric cell as we approach the critical point. This phenomenon, which contrasts with the *critical slowing down of diffusion*, is called **critical acceleration** or **"critical speeding up"**.

A temperature gradient that is resorbed in a cell results in density gradients, which, in turn, will be resorbed, but, this time, over a longer timescale, which is of the order of the diffusion time (or convection time, if convection is present). Heat transport thus takes place on the piston effect scale. However, it is necessary to wait for a period of time that is of the order of the diffusion time for the fluid to once again become homogeneous at equilibrium.

The numerical simulation results presented by Zappoli *et al.* (1996) made it possible to reject the false hypothesis, according to which the PE (piston effect) would disappear in the presence of gravity.

Indeed, this study showed that the PE was not inhibited by natural convection, but, on the contrary, is the fastest temperature relaxation mechanism, faster than convection. However, another experimental study conducted on the ground (Beysens 1992) showed that a non-insulated critical fluid cell undergoes almost no heating, compared to what occurs in microgravity. The analysis of the processes involved in this configuration was carried out by Jounet *et al.* (1999). Note that the piston effect has been the subject of experimental observations (Bonetti *et al.* 1994) (Figure 4.5).

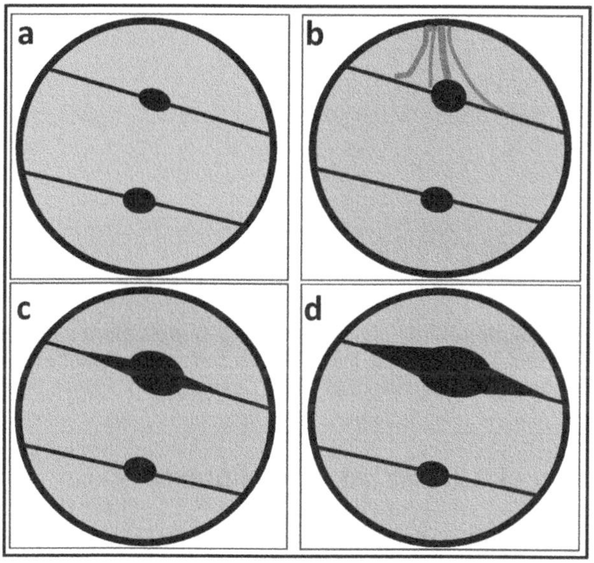

Figure 4.5. *Evolution of a thermal boundary layer around a heating thermistor (CO_2 to $\rho = \rho_c$). (a) Before heating. View of the cell with the two thermistors. The cell has a diameter of 11.6 mm. The two spheres are the thermistors. The lines are interference fringes. (b) Experiment under terrestrial gravity ($T = T_c + 16$ K, $t = 400$ ms after the heating starts). The convective plume can be observed. (c) Experiment in microgravity ($T = T_c + 16.8$ K, $t = 400$ ms, after the heating starts). (d) Same as in (c) at $t = 2$ s (schematic diagram redrawn from (Bonetti* et al. *1994, Figure 1))*

Flow calculations were made with a cavity of unit aspect ratio with length L = 10 mm, filled with supercritical CO_2. The bottom wall is heated linearly as a function of time, by 10 mK, for 1 s. The fluid is initially at 1 K from the critical point.

Figure 4.6 gives the temperature field under specified conditions. Thermal plumes can be observed near the hot and cold walls. The upper layer becomes unstable, testifying to the existence of a cold piston effect, with a contraction boundary layer along the upper wall placed at the initial temperature. The temperature is quite uniform between the walls (Amiroudine *et al.* 2001).

Figure 4.6. *Temperature field at t = 8.8 s with a temperature difference between the two walls of ΔT = 10 mK at 1 K from the critical point (schematic diagram redrawn from (Amiroudine* et al. *2001, extract from Figure 5, p. 133))*

4.4. Expansion of a "drop" at critical pressure

In the case of pure substances at supercritical pressures, there are no distinct liquid and vapor phases: the entire domain is a single phase. The relaxation in the density of a dense liquid pocket in a warmer and isobaric atmosphere is, nevertheless, similar to the process by which a subcritical liquid droplet vaporizes, under the condition that the thermal diffusivity of the liquid pocket is lower than that of the surrounding atmosphere. A zone with high density and diffusion time gradients insulates the dense pocket from the surrounding atmosphere (Sanchez-Tarifa *et al.* 1972; Préau *et al.* 2004). The dense pocket is thus called a supercritical "droplet", and the area with a high density gradient is compared to a thick "interface".

For a critical liquid pocket (critical density, critical temperature) introduced into a (hotter) atmosphere at critical pressure, the aforementioned diffusivity condition is valid for all temperatures of the atmosphere, due to the critical divergence of diffusivity in the critical pocket. At clearly supercritical pressures, this diffusivity state is more restrictive and, in ideal fluid conditions, requires $T_\infty >> T_i$, with T_∞ being the temperature of the atmosphere and T_i the initial temperature of the dense liquid pocket. Since the "interface" is thick, two limits can be chosen for the definition of the reference border of the supercritical "droplet": the first consists of choosing an isochor ρ_m located at the end of the high-density-gradient zone in the fluid pocket. We can observe that this surface, located at the end of the zone with the high-density gradient, is the site of maximal unit mass flow $r = r_m$. The second limit consists of the reference isochor ρ_0 at the start of the zone with the high-density gradient $r = r_0$. Between these two limits (the transition zone), we use a new non-dimensioned variable $X = (r - r_0)/(r_m - r_0)$, obtained by dividing the distances by the thickness of this "interface". The evolution is considered to be almost stationary with this new abscissa. For $r < r_0$ (the internal zone), the parameters are constant. For $r > r_m$, the evolution is almost stationary within the natural coordinates. These considerations make it possible to find an analytical solution. Let D_m be the diameter of the domain bounded by the isochor ρ_m and D_0 be the diameter of the domain bounded by the isochor ρ_0.

It is shown that the diameter D_m satisfies the "D²" law: $D_m^2 = 1 - \tau$, with τ dimensionless time, the ratio between the time and the life span of the drop. The diameter D_0 satisfies (see Figure 4.7):

$$D_o^\beta - \beta D_o = (1-\beta)\sqrt{1-\tau} \qquad [4.10]$$

The parameter β is of the order of unity at slightly subcritical or supercritical pressures. The D^2 law is not satisfied for D_0 at these pressures, as was demonstrated by the previous relationship. When pressure becomes clearly subcritical, β becomes large.

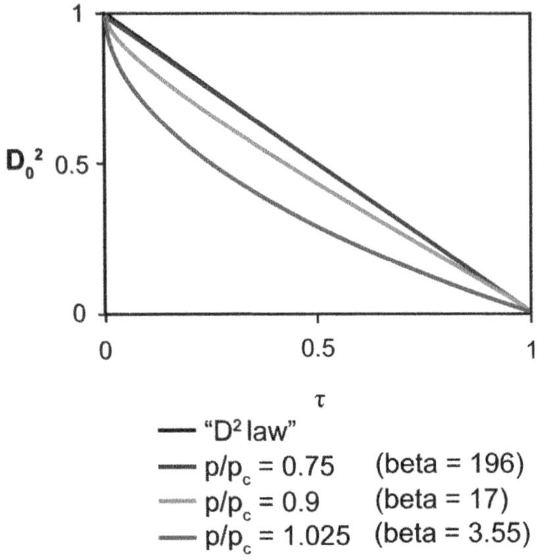

Figure 4.7. *Square of the diameter of the droplet as a function of reduced time τ for different pressures. p_c is the critical pressure. For a color version of this figure, see www.iste.co.uk/prudhomme/fluid1.zip*

Since the diameter D_o is dimensioned by the initial diameter of the droplet $D_0(\tau)\leq 1$ and, for large values of β, the earlier equation becomes, for a first-order equation: $D_o^2 = 1 - \tau$. We thus once again arrive at the D^2 law at clearly subcritical pressures.

The numerical solution, based on a finite volume discretization method for a symmetrical, spherical geometry, uses the SIMPLER algorithm to solve filtered equations of the system. The results of the analytical resolution solution are in very good agreement with those obtained numerically.

NOTE.– It would be interesting to define effective surface tension in supercritical conditions. Indeed, phenomena similar to those observed in miscible fluids leading to effective surface tension also appear in pure fluids when high-temperature gradients are produced (Gatignol and Prud'homme 2001). This is illustrated in Figure 4.8, in the case of a cold drop falling into the same liquid at a higher temperature (Kojima *et al.* 1984).

A slightly different problem is brought into play, notably in the injection zone of cryogenic engines. The isobaric vaporization, at supercritical pressure, of a cold "droplet" in a hot atmosphere must be considered. The droplet is initially at a subcritical temperature. Outside the near-critical zone, there is no singularity of thermal conductivity, and thus there is a period where the drop heats up, as occurs in the case of a liquid immersed into a perfect gas; however, here, there is an equation of state for a real gas.

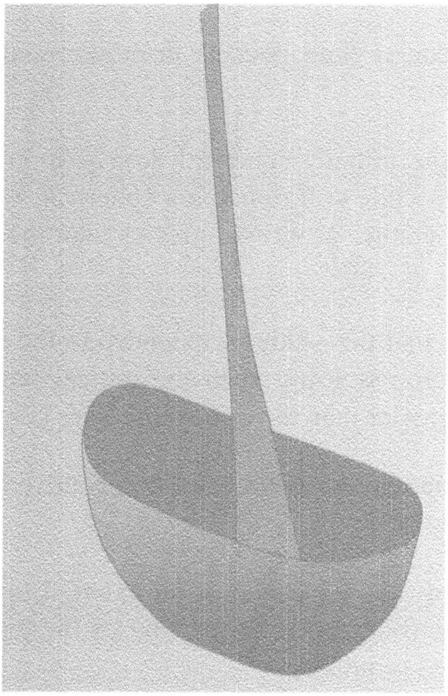

Figure 4.8. *A thermal drop. The cold drop at -7°C falling into the same liquid, initially at 21°C (schematic diagram redrawn from (Kojima* et al. *1984))*

4.5. Behavior of a pocket of supercritical fluid immersed into a high-temperature environment

The case of a drop of pure fluid in the critical state immersed in an isobaric environment with lower density, which was the subject of Preau's work at LMM[1], made it possible to analytically and numerically establish the

1 LMM: *Laboratoire de Modélisation en Mécanique* (Mechanical Modeling Laboratory).

evaporation law; the consequences of this law are yet to be explored, and other cases of evaporation, where the initial drop is at critical pressure but at a temperature that is lower than the critical temperature, are also yet to be studied.

In certain combustion chambers, drops of liquid (fuel or oxidant) are injected into a very hot atmosphere at a pressure greater than the critical pressure of the liquid under consideration.

These drops appear as fluid pockets in motion and expansion, with no phase change interface.

The case where the initially spherical drop is in critical conditions and expands isobarically, while still remaining spherical, within the same fluid, at rest and at a temperature at infinite greater than the critical temperature, was studied (Préau *et al.* 2004).

Here, we propose that this study be extended to the case where the initial temperature of the drop is lower than the critical temperature and where ambient pressure is greater than the critical pressure (Figure 4.9).

This situation is quite representative of cryogenic engines (e.g. an Ariane rocket) and diesel engines.

However, in reality the existence of an external turbulent flow and the absence of surface tension cause the peeling of the pocket and lead to the rupture of its spherical symmetry.

Nevertheless, we will assume here that the pocket remains spherical, not only for simplification but also to establish laws of evolution (D'Almeida and Prud'homme 2009), which may be introduced into more complex models that include turbulence. The numerical part of the study was carried out using an existing code provided by Jalil Ouazzani. For the analytical part of the study, asymptotic expansions are recommended. An analytical and numerical study was carried out in this manner by Sagna and D'Almeida (2013), by postulating additional hypotheses about the thermodynamic properties of the gaseous phase, in order to bring back the law in conventional D^2 (Sagna and D'Almeida 2014), which is not possible in the critical case leading to equation [4.10].

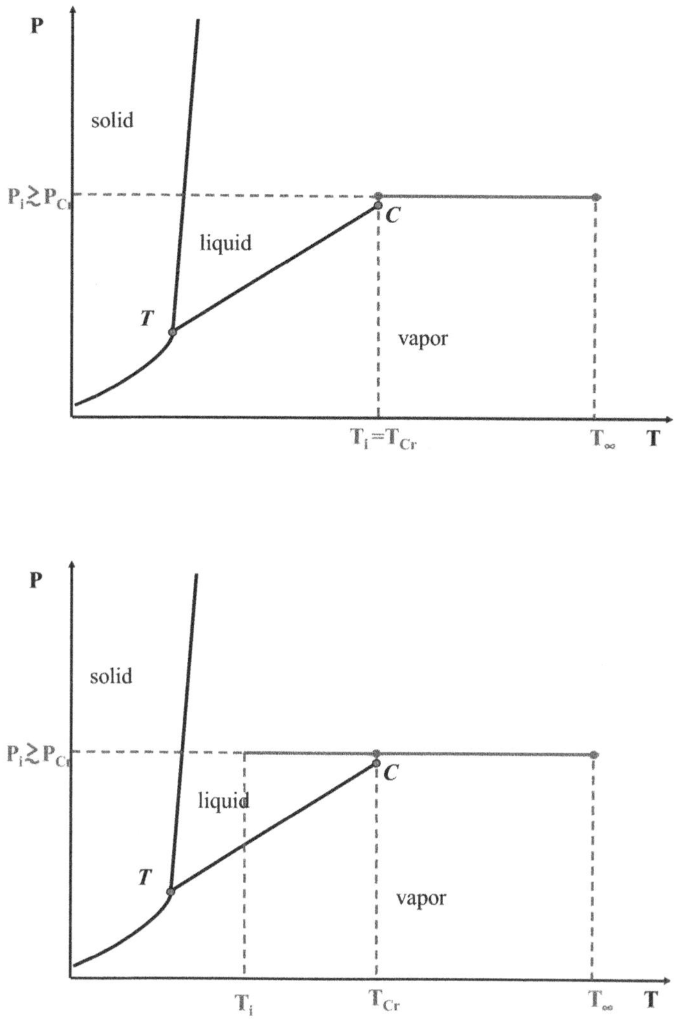

Figure 4.9. *Temperature–pressure diagram of a pure body; T: triple point, C: critical point, (T_i, p_i): initial conditions, (T_∞, p_i): final conditions. a) Near-critical drop in the initial state (Préau* et al. *2004); b) cold drop in the initial state (Sagna and D'Almeida 2014). For a color version of this figure, see www.iste.co.uk/prudhomme/fluid1.zip*

It must be noted that the problem of heating pockets of dense gases in flows has already been studied numerically. Figure 4.10 shows the deformations undergone by such a pocket.

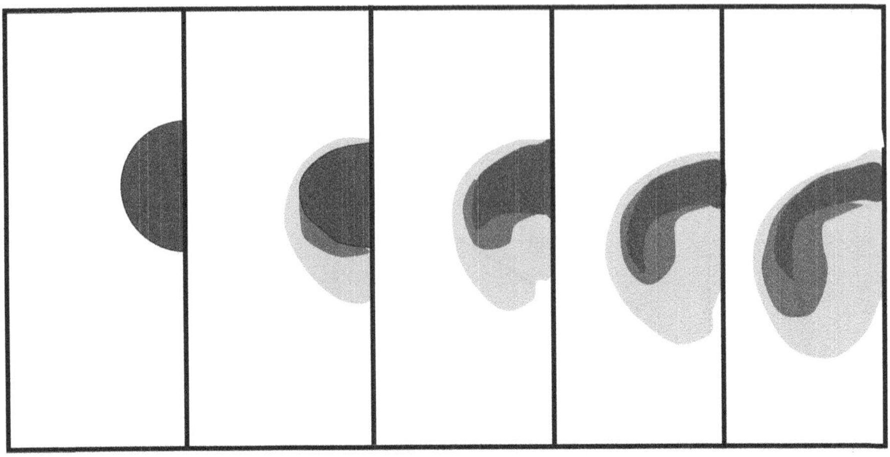

Figure 4.10. *Heating and deformation of a dense pocket over its life span. The range of hues is relative to the temperature field. Reynolds injection number: Re = 100, Prandtl number: Pr = 1 and density ratio = 15. Dimensional injection temperature: 100 K, ambient temperature:* T_∞ *= 1500 K, initial radius of the pocket:* a_0 *= 50 μm (schematic diagram redrawn from (Kojima* et al. *1984))*

4.6. Boiling near the critical point[2]

Bubble growth experiments were carried out with the DECLIC[3] instrument in the International Space Station. DECLIC is dedicated to the study of solidification in transparent media and the study of critical fluids.

NOTE.– One advantage of conducting experiments close to the critical point is the relative slowness of phenomena, which makes observation much easier.

The hyper-compressibility of fluids in the critical region allows extremely rapid adiabatic transportation of heat, which has been shown in zero gravity (Garrabos 1998) and christened the "piston effect": the thermal boundary layer close to the heating wall expands and compresses, heating the rest of the fluid at the speed of sound. New experiments were carried out using the

2 According to Yves Garrabos, CNRS Talence, and Daniel Beysens, CEA/ESPCI Paris.

3 DECLIC: *Dispositif d'Étude de la Croissance et des Liquides Critiques* (Device for the Study of Growth and Critical Liquids).

ALICE2 experimental devices in the MIR station, and these demonstrated that this heat transport mechanism may lead to paradoxical results (Wunenburger *et al.* 2000), seemingly violating the second law of thermodynamics, as has been cited (Sincell 2000): when the two phases (gas and liquid) of a fluid (SF_6) are heated in zero gravity, the gas may indeed become *hotter* than the heating wall!

This overheating is at its maximum at 5°C from the critical point, where it is greater than 20% (Figure 4.10). This paradox may be removed if it is considered that thermal transfer and mechanical transfer take place: the liquid layer, expanded by the heating of the cell body, compresses the vapor more than the liquid. The heat-compressible nature of the piston effect thus explains this apparent violation of the thermodynamic law.

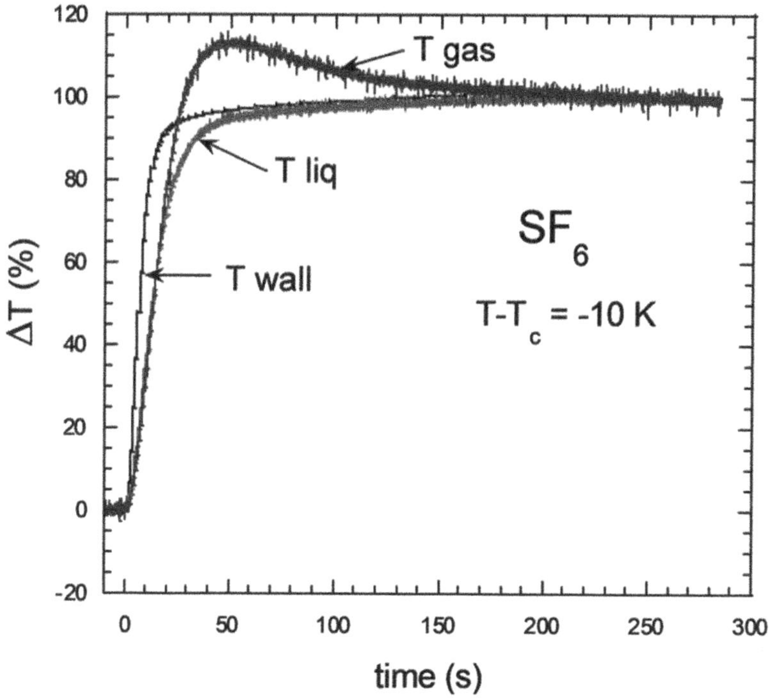

Figure 4.11. *Overheating observed during a thermal quenching of 0.1°C in an SF_6 cell that was initially at 10°C below its critical point: the temperature of the gas is greater than that of the heating wall (taken from work on DECLIC). For a color version of this figure, see www.iste.co.uk/prudhomme/fluid1.zip*

This anomaly is only visible if the gas bubble is far from the cell wall. If this is not the case, other phenomena are observed in relation to the boiling crisis that thermicians are familiar with, when the heat flow is so strong that the liquid evaporates and forms a vapor film on the wall, preventing the transfer of heat.

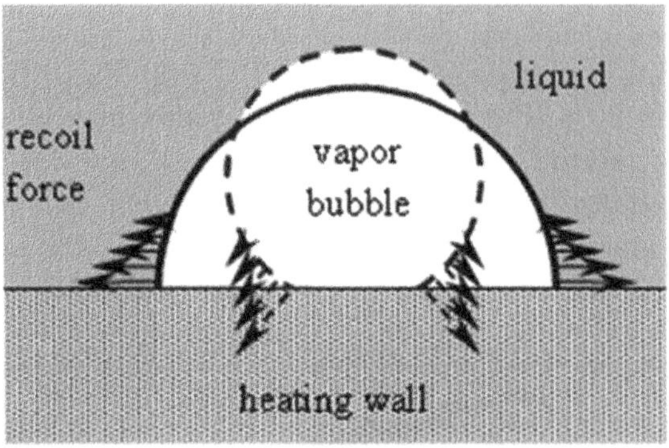

Figure 4.12. *The heat bubble stretching over a heating wall, under the action of the vapor recoil force (taken from works on DECLIC). For a color version of this figure, see www.iste.co.uk/prudhomme/fluid1.zip*

This phenomenon was interpreted (Nikolayev and Beysens 1999) using the dynamics of the solid–liquid–vapor triple line, where the evaporation thrust produces a force (recoil force), which draws the contact line towards the liquid phase and spreads the vapor over the wall (Figure 4.11). In a cell where the gas bubble is now compressed by the walls, the gas *spreading* over the wall (Garrabos *et al.* 2001) is observed (Figure 4.12). This is a true *boiling crisis*, visible here with low thermal flow, as the absence of gravity and the proximity to the critical point make the forces of gravity and the latent heat of vaporization negligible. This transition, through pulling down the contact line, is an experimental demonstration of ideas developed in theoretical and numerical studies (Nikolayev *et al.* 2000), where this mechanism was proposed in order to take into account the boiling crisis (Figure 4.13).

An update on these questions related to microgravity has recently been published (Nikolayev *et al.* 2016).

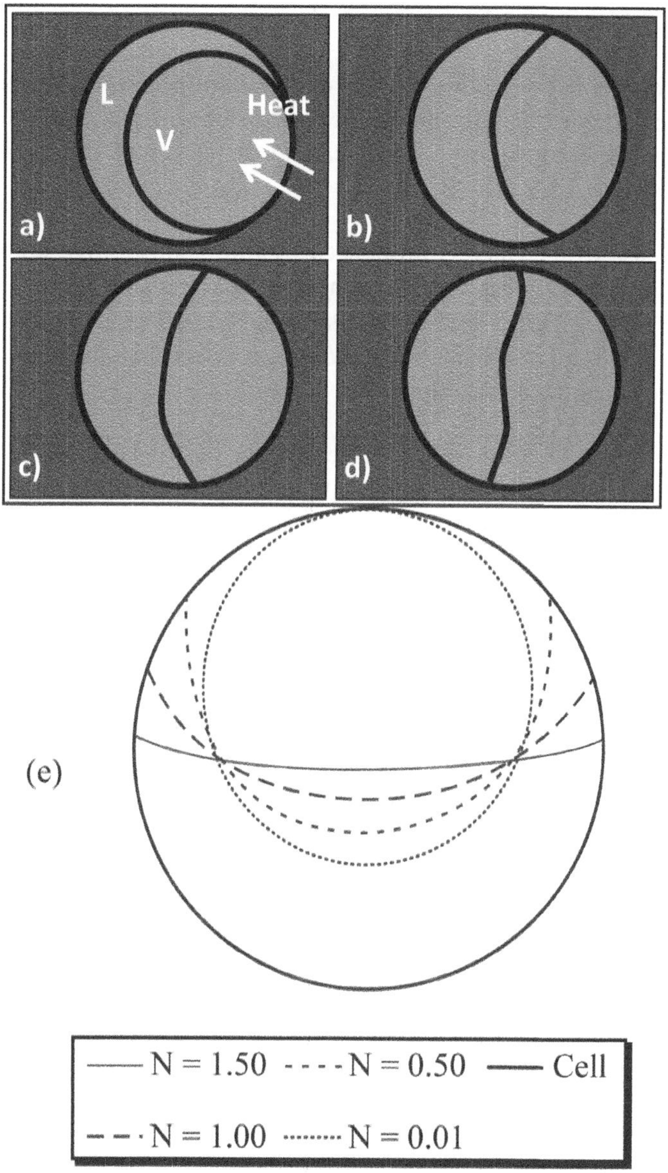

Figure 4.13. *Deformation of the liquid–vapor interface under the action of the recoil force. (a–d) SF_6 fluid in zero gravity, subject to a heat flow. (a) $T - Tc = 10.4$ K, (b) $T - T_c = 0.25$ K, (c) $T - T_c = 0.15$ K, (d) $T - T_c \approx 0$. (e) 2D simulation: the parameter N undergoes variations corresponding to the difference $(T - T_c)/T_c$. Vapor volume fraction: 0.5. The actual angle of contact is zero (schematic diagram redrawn from (Nikolayev* et al. *2000))*

4.7. Conclusion

This chapter began with a review of the thermodynamics of critical and supercritical fluids, because the properties of these fluids of fluids are very particular. Their constitutive laws lead to an effect of *divergence at the critical point.*

Every case of interface is influenced by these constitutive laws: whether this is the *piston effect*, which concerns heat transfer at the wall, the *expansion of a drop* at critical pressure, where phenomena occur on the surface of the drop, or whether this is *boiling*, where there is the recoil action of the vapor near the contact line with the wall.

Being in the vicinity of the critical point has a major effect on these interface phenomena, whose mathematical analysis, numerical modeling and experimental observation all offer unique features, as we have seen.

On an experimental level, there was a systematic use of microgravity and the quality required (in terms of precision and duration) of this microgravity led scientists to opt for satellite experimentation, with the help of evolved diagnostics. Therefore the results that are obtained are due to special efforts of intensity and duration. This would not have been possible without the structural, scientific and financial support from the CNES, CEA, CNRS and university laboratories. International cooperation has played a key role, not only on the scientific level, but also in regards to satellite launch vehicles.

4.8. References

Amiroudine, S., Bontoux, P., Laroudé, P., Gilly, B., Zappoli, B. (2001). Direct numerical simulation of instabilities in a two-dimensional near-critical fluid layer heated from below. *J. Fluid Mech.*, 442, 119.

Arias-Zugasti, M., Garcia-Ibarra, P.L., Castillo, J.L. (1999). Droplet vaporization at critical conditions: Long-time convective-diffusive profiles along the critical isobar. *Phys. Rev. E*, 60(3), 2930–2941.

Bachaud, C. (1951). La variation de la tension superficielle près du point critique : rôle des densités du liquide et de la vapeur saturante. *J. Phys. Radium*, 12(8), 825–826.

Beysens, D. (1992). Que peut-on apprendre sur les phénomènes de transport d’une étude en microgravité. *Ann. Chim. Fr.*, 17, 45–53.

Beysens, D. (2001). Near-critical fluids under micro gravity: Highlights and perspectives for Europe. *J. Phys. IV France*, 11.

Bonetti, M., Perrot, F., Beysens, D. (1994). Fast thermalization in supercritical fluids. *Phys. Rev.*, 49(6).

Boukari, H., Shaumeyer, J.N,, Briggs, M.E., Gammon, R.W. (1990). Critical speeding up in pure fluids. *Phys. Rev. A*, 41, 2260.

D'Almeida, A. and Prud'homme, R. (2009). Evaporation de gouttes lois de récession du rayon. *19ème congrès français de mécanique*, Marseille, France, 24–28 August.

Garrabos, Y., Le Neindre, B., Subra, P., Cansel, F., Pommier, C. (1992). Fluides critiques et gravité, fluides supercritiques et matériaux. *Fluides, Matériaux et Microgravité, Ann. Chim. Fr.*, 17, 55–90.

Garrabos, Y., Lecoutre-Chabot, C., Hegseth, J., Nikolayev, V., Beysens, D., Delville, J.-P. (2001). Gas spreading on a heated wall wetted by liquid. *Phys. Rev. E*, 64, 1–10.

Gatignol, R. and Prud'homme, R. (2001). *Mechanical and Thermodynamical Modeling of Fluid Interfaces*. World Scientific, Singapore.

Jounet, A., Zappoli, B., Mojtabi, A. (1999). Simulation numérique de l'effet piston avec prise en compte des échanges thermiques pariétaux. *C.R.A.S. Paris*, 327 (IIb), 991–996.

Kojima, M., Hinch, E.J., Acrivos, A. (1984). The formation and expansion of a toroidal drop moving in a viscous fluid. *Phys. Fluids*, 27(1), 19–32.

Luettmer-Strathmann, L., Sengers, J.V., Olchowi, G.A. (1995). Non-asymptotic critical behavior of the transport properties of fluids. *J. Chem. Phys.*, 103(17), 7482–7501.

Mezger, E. (1946). Loi de variation de la tension superficielle avec la température. *J. Phys. Radium*, 7(10), 303–309.

Nikolayev, V. and Beysens, D. (1999). Boiling crisis and nonequilibrium wetting transition. *Europhys. Lett.*, 47, 345–351.

Nikolayev, V., Beysens, D., Lagier, G.-L., Hegseth, J. (2000). Growth of a dry spot under a vapor bubble at high heat flux and high pressure. *Int. J. Heat and Mass Transfer*, 44(18), 3499–3511.

Nikolayev, V.S., Garrabos, Y., Lecoutre, C., Pichavant, G., Chatain, D., Beysens, D. (2016). Evaporation condensation-induced bubble motion after temperature gradient set-up. *Comptes rendus – mécanique*, 345(1), 35–46.

Nitsche, K. and Straub, J. (1987). The critical "hump" of C_v under microgravity, result from the D-1 Spacelab experiment "Wärmekapazität". *Proceedings of 6th European Symposium on Material Science Under Microgravity Conditions*, 109–116, Bordeaux, France, 2–5 December.

Onuki, A., Hong, H., Ferreli, R.A. (1990). Past adiabatic equilibration in a single component fluid near the liquid-vapor critical point. *Phys. Rev. A*, 41(2256).

Prausnitz, J., Rüdiger, M., Lichtenthaler, N., Gomes de Azevedo, E. (1999). *Molecular Thermodynamics of Fluid-Phase Equilibria*, 3rd edition. Prentice Hall, Upper Saddle River.

Préau, S., Prud'homme, R., Ouazzani, J., Zappoli, B. (2004). Supercritical density relaxation as a new approach of droplet vaporization. *Phys. Fluids*, 16, 4075–4087.

Prud'homme, R. (1988). *Fluides hétérogènes et réactifs : écoulements et transferts*. Springer-Verlag, Berlin, Heidelberg.

Prud'homme, R. (2016). Bernard Zappoli's contribution to the French scientific life. *Comptes rendus – mécanique*, 345(1), 3–10.

Reid, R.C., Prausnitz, J.M., Poling, B.E. (1987). *The Properties of Gases and Liquids*, 4th edition. McGraw-Hill, Boston.

Sagna, K. and D'Almeida, A. (2013). A study of droplet evaporation. *Am. J. Modern Phys.*, 2(2), 71–76.

Sagna, K. and D'Almeida, A. (2014). Supercritical evaporation of a drop. *Int. J. Res. Eng. and Sci.*, 2(7), 41–47.

Sanchez-Tarifa, C., Crespo, A., Fraga, E.A. (1972). Theoretical model for the combustion of droplets in super-critical conditions and gas pockets. *Astron. Acta*, 17, 685–692.

Sengers, J.V. (1985). Transport properties of fluids near critical points. *International Journal of Thermophysics*, 6, 203.

Sengers, J.V. and Levelt Sengers, J.M.H. (1978). Critical phenomena in classical fluids. In *Progress in Liquid Physics*, Croxton, C. (ed.). Wiley, Chichester.

Sincell, M. (2000). Backward heat flow bends the law a bit. *Science*, 288(5467), 789–791.

Wunenburger, R., Garrabos, Y., Chabot, C., Beysens, D., Hegseth, J. (2000). Thermalization of a two-phase fluid in low gravity: Heat transfer from cold to hot. *Phys. Rev. Lett.*, 84, 4100–4103.

Zappoli, B., Bailly, D., Garrabos, Y., Le Neindre, B., Guenoun, P., Beysens, D. (1990). Anomalous heat transport by the piston effect in supercritical fluids under zero gravity. *Phys. Rev. A*, 41(4), 2264–2267.

Zappoli, B., Amiroudine, S., Carlès, P., Ouazzani, J. (1996). Thermoacoustic and bouncy-driven transport in a square side-heated cavity filled with a near-critical fluid. *J. Fluid Mech.*, 316, 53–72.

Zappoli, B., Beysens, D., Garrabos, Y. (2015). *Heat Transfers and Related Effects in Supercritical Fluids*. Springer Netherlands, Dordrecht.

5

Shear-Induced Anomalies in the Brownian Motion of Particles in Strongly Fluctuating Near-Critical Mixtures

The Brownian motion of colloids in a fluid is influenced by its local viscosity through their drag coefficient. In near-critical binary mixtures, viscosity increases when nearing the critical point as an effect of increasing concentration fluctuations. It happens that the local shear flow around the moving particles can modify the local viscosity and saturate its enhancement when fluctuations are large enough to be deformed by shear. Using photon-beating spectroscopy, it is observed that latex beads immersed in isobutyric acid and water exhibit a slowdown of their Brownian motion due to the fluctuation-induced critical enhancement of viscosity. However, as temperature is brought close enough to the CP value, local viscosity saturates as an effect of shear flow and colloid motion ceases to slow down, reaching a plateau. The temperature dependence of the particle self-diffusion coefficient is seen to agree with a recent estimation by Fujitani (2021) of the drag coefficient of the particle using a Langevin equation.

5.1. Introduction

Investigation of the Brownian motion of small particles (colloids) immersed in a liquid has a long history, going back to the works of Einstein and Perrin (Einstein 1905; Perrin 1909) at the early beginning of the 20th century. The use of modern techniques (fast videos, optical tweezers) with high spatial and temporal resolution has recently given rise to the precise determination of instantaneous velocity (Li *et al.* 2010; Huang *et al.* 2011;

Chapter written by Daniel BEYSENS.

Kheifets *et al.* 2014). The hydrodynamic coupling between the colloid and the fluid has been seen to dominate the dynamics of velocity fluctuations. In particular, the equipartition theorem is verified only if we consider the "added mass" of fluid accounting for the inertia of the displaced fluid in the colloid motion (Zwanzig and Bixon 1975).

Brownian motion can thus be affected by the local hydrodynamics around the colloids. It happens that hydrodynamic coupling can be magnified when nearing a critical point where the local fluctuations of the order parameter affect the fluid resistance to flow, resulting in viscosity enhancement (Stanley 1987; Onuki 2002). When a colloid is immersed in a binary liquid near its liquid–liquid critical point, its Brownian motion can be affected by the fluctuation-induced enhancement of viscosity as a result. However, the shear flow around the colloid can in turn reduce the local concentration fluctuations, resulting in saturation of the friction (see section 5.2.2). This complex hydrodynamic process disturbs the Brownian motion of colloids, as shown here by the result of photon-beating light scattering experiments performed in a mixture of isobutyric acid and water above its critical point, where it is homogeneous.

This chapter is organized as follows. The fundamentals of critical point phenomena and effects of a shear flow are discussed first. Then, the experimental methods are described and the results are presented and considered. The basics of the photon-beating technique are given in section 5.7.

5.2. Theoretical background

5.2.1. *Miscibility critical point in binary mixtures*

The vicinity of second-order phase transition (critical point) exhibits a number of universal, scaled behaviors (Stanley 1987; Zappoli *et al.* 2015). Fluids near their liquid–vapor critical point, and partially miscible binary mixtures near their liquid–liquid critical point (the upper point in the temperature–density or concentration phase diagram), exhibit the same features as they belong to the same universality class. The representative class is the magnetic Ising model with magnetization M as the order parameter. In binary fluids, M is the concentration difference c - c_c, with c the concentration of one of the components and c_c the critical concentration. This class is characterized by the space dimensionality D = 3 and order parameter dimensionality d = 1. An important aspect of the critical region is that most of the anomalies of the thermodynamic and transport properties can be set in the form of scaled, universal power laws with respect to the critical point coordinates. For instance, the correlation length of concentration fluctuations, ξ, asymptotically diverges with the reduced temperature $\varepsilon = \frac{T-T_c}{T_c}$ (here T is the absolute temperature and T_c is the critical temperature) along the $c = c_c$ thermodynamic path, such that

$$\xi = \xi_0 \epsilon^{-\nu} \qquad [5.1]$$

Here, ν is a universal exponent and ξ_0 is a system-dependent critical amplitude. The value of critical exponents depends on D. Above an upper space critical dimensionality D_c = 4, the effect of fluctuations on divergence can be neglected, mean-field approach is relevant and ν = 1/2. For the class of fluids where the space dimensionality $D = 3 < D_c$, fluctuations have to be accounted for, which renormalizes the exponent value to ν = 0.630 (Stanley 1987; Zappoli *et al.* 2015).

The lifetime of concentration fluctuation corresponds to the diffusion of a component on length scale ξ. The fluctuation indeed vanishes by a diffusion process, with a diffusion coefficient D_ξ, which can be related to the Brownian diffusion of a cluster of size ξ (Stanley 1987; Onuki 2002; Zappoli *et al.* 2015). With k_B the Boltzmann constant and η the liquid mixture shear viscosity, the self-diffusion constant therefore reads as

$$D_\xi = \frac{k_B T}{6\pi\eta\xi} \qquad [5.2]$$

Diffusion on length ξ occurs on time τ such that

$$\tau = D_\xi^{-1}\xi^2 = \left(\frac{6\pi\eta}{k_B T}\right)\xi^3 = \left(\frac{6\pi\eta\xi_0^3}{k_B T}\right)\epsilon^{-3\nu} \qquad [5.3]$$

Shear viscosity, defined as the ratio of the shear stress to an infinitesimally small velocity gradient or shear flow S – the so-called Newton's law for fluids – is also affected by the coupling with critical fluctuations (Onuki 2002; Zappoli *et al.* 2015). Viscosity exhibits a (weak) divergence such that

$$\eta^{(0)} = \eta_B(T)\epsilon^{-Y} \quad [5.4]$$

Here, the critical exponent is $Y = 0.04$ and $\eta_B(T)$ is a thermally activated regular (background) contribution of Arrhenius form, corresponding to an energy barrier to be overcome for molecules to move. With E the activation energy and η_0 an amplitude,

$$\eta_B(T) = \eta_0 \exp(\frac{E}{k_B T}) \quad [5.5]$$

5.2.2. *Effect of shear flow*

Spatial fluctuations near the critical point can be strongly deformed by external influences as hydrodynamics. Shear flow alters the spatial fluctuations (Beysens and Gbadamassi 1979; Onuki 2002) when the shear-induced fluctuation deformation, l, during the fluctuation lifetime τ, becomes larger than the fluctuation correlation length ξ, i.e. $l = S\xi\tau > \xi$ (Figure 5.1), or

$$S\tau > 1 \quad [5.6]$$

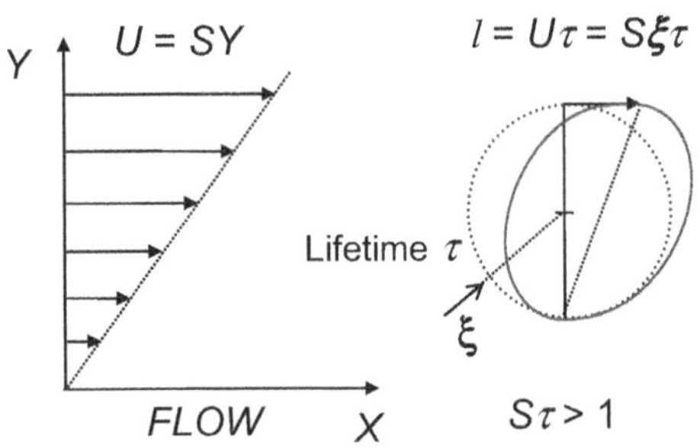

Figure 5.1. *Schematics of a symmetric fluctuation (dotted) deformed by shear flow (continuous) when its lifetime τ is large enough such that the length scale of deformation $l = S\xi\tau > \xi$ or $S\tau > 1$. U: liquid velocity; for other symbols, see text. For a color version of this figure, see www.iste.co.uk/prudhomme/fluid1.zip*

This means that the fluctuation lifetime is long enough to "feel" the shear. The above condition equation [5.6], when considering the fluctuation lifetime expression equation [5.3], corresponds to having a cross-over correlation length, ξ_C, above which fluctuations become strongly deformed by shear:

$$\xi_C = \left(\frac{k_B T}{6\pi\eta}\right)^{1/3} S^{-1/3} \qquad [5.7]$$

When nearing a critical point on the thermodynamic path $c = c_c$ by changing temperature, the well-known "critical slowing down", where τ increases (equation [5.3]), ensures that the condition equation [5.6] will eventually be fulfilled. From this equation [5.6] derives a condition on temperature, separating on both sides of $\varepsilon = \varepsilon_C$ two regions of low shear ($\varepsilon > \varepsilon_C$ and high shear ($\varepsilon < \varepsilon_C$), where

$$\varepsilon_C = \left(\frac{6\pi\eta\xi_0^3 S}{k_B T}\right)^{1/3\nu} \qquad [5.8]$$

The reduction of fluctuations in the high-shear region is such that mean-field behavior, where fluctuations can be neglected, becomes relevant near the CP (Oxtoby 1975; Beysens and Gbadamassi 1979; Onuki 2002). In other words, the upper critical dimensionality is lowered (to $D_c = 2.4$ (Onuki 2002)) below the space dimensionality D=3. In fact, viscosity in the mean-field approximation does not show any more a divergence and levels off (Oxtoby 1975; Onuki 2002). It results in a suppression of the viscosity divergence.

In order to account for this behavior, we can represent the viscosity by the following simplified function (Fukitani 2021):

$$\eta = \begin{cases} \eta_B(T)\epsilon^{-Y} & \text{for } \epsilon > \epsilon_C \\ \eta_B(T){\varepsilon_C}^{-Y} & \text{for } \epsilon < \epsilon_C \end{cases} \qquad [5.9]$$

In the high-shear region $\epsilon < \epsilon_C$, the viscosity reduction $\Delta\eta$ with respect to equation [5.4] behavior can thus be expressed in terms of relative temperature $\tilde{\epsilon}_C = \varepsilon_C/\varepsilon$ such that

$$\Delta\eta = \eta_B(T)\epsilon^{-Y} - \eta_B(T){\varepsilon_C}^{-Y} = \eta_B(T)\epsilon^{-Y}(\tilde{\epsilon}_C^{-Y} - 1)\Theta(\tilde{\varepsilon}_C - 1) \qquad [5.10]$$

Here, the step function $\Theta(x)$ is introduced, with $\Theta(x) = 0$ for $x < 0$ (low shear) and $\Theta(x) = 1$ for $x > 0$ (high shear). The quantity

$$\frac{\Delta\eta}{\eta^{(0)}} = \tilde{\epsilon}_C^{-Y} - 1 \qquad [5.11]$$

thus represents, in the high-shear region where $\Theta = 1$, the relative modification of viscosity due to shear.

5.2.3. *Colloid Brownian motion and shear flow*

When a particle (colloid) is immersed in a liquid, it undergoes Brownian motion with self-diffusion constant D (Einstein–Stokes). Assuming that viscosity is uniform and flow laminar (low Reynolds number), D is expressed as:

$$D = \frac{k_B T}{6\pi\eta a} \qquad [5.12]$$

Brownian motion generates a flow field around the particle. If the liquid is a near-critical binary liquid, the local shear flow will locally modify the viscosity according to the distance to the CP according to equation [5.11], making the viscosity inhomogeneous (Figure 5.2).

The mean value of the colloid velocity, $\overline{U}$, can be inferred from the equipartition theorem. For each Cartesian component of velocity, it can be written as:

$$\overline{U} = \sqrt{\frac{k_B T}{m^*}} \qquad [5.13]$$

In a dense liquid, the inertial mass of an object has to be modified by the inertia of the displaced fluid (the so-called added mass m_a), which is half of the mass of the displaced fluid (Zwanzig and Bixon 1975). For a sphere of radius a immersed in a fluid with density ρ_f, $m_a = \frac{2\pi}{3} a^3 \rho_f$. With the sphere mass being $m = \frac{4\pi}{3} a^3 \rho_s$, where ρ_s is the sphere density, the effective mass m^* reads as

$$m^* = \frac{2\pi}{3} a^3 (\rho_f + 2\rho_s) \qquad [5.14]$$

When dealing with spherical colloids in Brownian motion, the Reynolds number is generally very low (see equation [5.18] and its evaluation), indicating that the flow is dominated by viscous effects around the sphere.

Stokes damping can thus be successfully assumed. Hydrodynamics can be considered stationary as it corresponds to a diffusive behavior over times longer than the viscous time $a^2\rho_f/\eta \approx 2.6$ ns for the colloids used here (see Table 5.1 for the numerical values). This time is much shorter than the binary liquid fluctuation time $\tau > 50$ ns (values from Table 5.1) in the studied temperature range.

The stationary velocity field around a sphere with velocity U at a small Reynolds number is axisymmetric and is well known (Happal and Brenner 1963). In polar coordinates (r, θ), the components of fluid velocity, u_r, along r, and u_θ, perpendicular to r, are

$$u_r(r,\theta) = -U\cos\theta\left[-\frac{3}{2}\frac{a}{r} + \frac{1}{2}\left(\frac{a}{r}\right)^3\right] \quad [5.15]$$

and

$$u_\theta(r,\theta) = -U\sin\theta\left[\frac{3}{4}\frac{a}{r} + \frac{1}{4}\left(\frac{a}{r}\right)^3\right] \quad [5.16]$$

The shear rate along r can be deduced by derivating equation [5.15] with respect to r for $r/a>1$. The maximum shear flow S corresponds to u_r at $\theta = 0$ and $r/a = \sqrt{2}$, giving $S_r = \left(\frac{du_r}{dr}\right)_{max} = \frac{3}{8}\frac{U}{a}$. The maximum shear rate perpendicular to r corresponds to u_θ at $\theta = \pi/2$ and $r = a$, giving $S_\theta = \left(\frac{du_\theta}{dr}\right)_{max} = \frac{3}{2}\frac{U}{a}$. We will thus consider in the following the mean value

$$S \approx \frac{\bar{U}}{a}. \quad [5.17]$$

With the numerical values of Table 5.1, we find $\bar{U} = 3.6\times10^{-2}$ ms^{-1} (equation [5.13]), from which it follows $S = 4.5 \times 10^5$ s^{-1}.

In order to look in more detail at how shear and viscosity enhancement is distributed around a particle with velocity $\bar{U}$, Fujitani (2021) calculated a map of the shear rate to calculate the dependence of the drag coefficient on the particle speed. Figure 5.2 shows how viscosity is modified at temperature threshold $\epsilon = \epsilon_c$ and below, $\epsilon < \epsilon_c$.

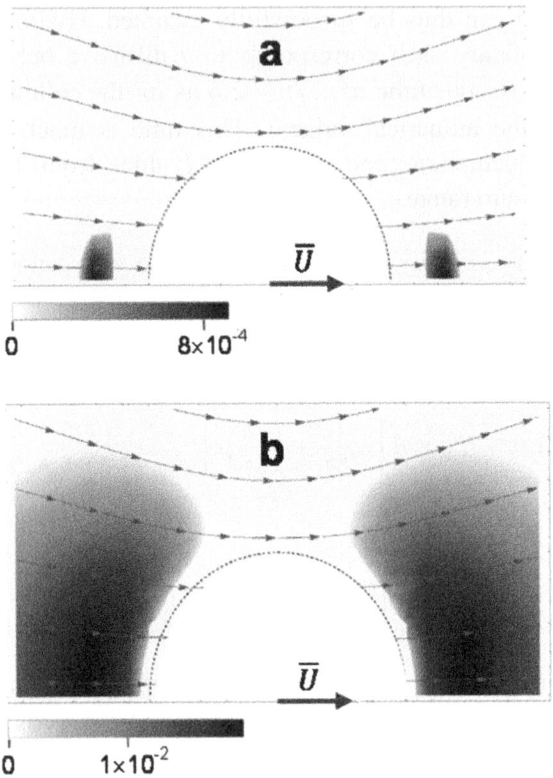

Figure 5.2. *Suppression of viscosity enhancement (colored zones) due to the shear flow induced by the Brownian motion of a particle. The particle is assumed to move translationally with mean velocity $\overline{U}$ (equation [5.13]). The curves outside the particle are the stream lines. Colors represents the viscosity variation $\Delta\eta/\eta^{(0)} = \tilde{\epsilon}_c^{-Y} - 1$ at (a) $\epsilon \approx \epsilon_c$ = 1.26 × 10⁻² and (b) $\epsilon = 8.064 \times 10^{-3} < \epsilon_c$ (adapted with permission from Fujitani (2021)). For a color version of this figure, see www.iste.co.uk/prudhomme/fluid1.zip*

5.3. Experiments and methods

The colloids considered here are latex beads from Dow Chemical whose characteristics are listed in Table 5.1. The latex bead suspensions are composed mainly of polymer particles and water, with small amounts of surfactant, sodium bicarbonate and potassium sulfate. A typical suspension contains the following: water > 69.0%, polymer 30.0%, surfactant 0.1–0.5%, inorganic salts 0.2%. A very dilute solution with water was added to the mixture of isobutyric acid and water. Photon-beating spectroscopy of beads (see section 5.7) far from T_c gives $a = 8.0 \times 10^{-8}$ m from equations [5.19] and [5.20]. It is implicitly assumed that the flow is dominated by viscosity around the colloids, as assessed by the Reynolds number <<1 (see equation [5.18]).

The binary mixture of isobutyric acid and water exhibits an upper critical point with a critical mass fraction of isobutyric acid $c_c = 0.3889$ and temperature $T_c = 301$ K (Toumi and Bouanz 2008). A precise value is not necessary as minute quantities of impurities can shift T_c. What really matters is the temperature difference $T - T_c$, with T_c experimentally determined for each sample. The refractive index at the critical point is $n_c = 1.36105$ for the laser wavelength $\lambda_0 = 632.8$ nm. With a mean temperature dependence $(dn/dT) \approx -2.5 \times 10^{-4}$ K^{-1} and a studied temperature range $T - T_c \approx 20$ K, the refractive index thermal variation $\Delta n < 5 \times 10^{-3}$ and can be neglected. In the following, the value $n = 1.361$ will be used (Table 5.1). Concerning density, its temperature variation is also negligible and the value at T_c, $\rho_f = 990$ kg.m^{-3} (Jacobs and Greer 1996; Toumi and Bouanz 2008) is reported in Table 5.1. Isobutyric acid of quality better than 99.99 mol% was used without further purification. Ultrapure water from a Merck Millipore purification device was used. The experimental isobutyric acid mass fraction used was 0.3887 ± 0.0002, close enough to c_c.

The mixture with colloids is set in a sealed glass sample, itself put in a thermostat with temperature regulation to within $\pm 10^{-3}$ K. The beam of a 10 mW He–Ne polarized laser is slightly focused in the sample. Collimated light scattered at $\alpha = 90°$ is recorded by a photomultiplier, whose homodyne amplified signal is then analyzed by a correlator (the basics of photon-beating spectroscopy are given in section 5.7). The critical temperature is determined within ± 5 mK by decreasing the temperature step by step and looking at the phase separation process on a screen in front of the sample, perpendicular to the laser beam. The value $T_c = (301.130 \pm 0.005)$ K was found. Note that the signal due to light scattered by colloids is more important than the signal due to the critical fluctuations of the mixture, preventing interference in the analysis.

								Shear viscosity	
	Critical temperature	*Correlation length amplitude*	*Refractive index*	*Incident wavelength*	*Fluid density*	*Bead density*	*Bead radius*	*Amplitude*	*Activation energy*
	T_c (K)	ξ_0 (nm)	n	λ_0 (nm)	ρ_f(kg.m^{-3})	ρ_s (kg.m^{-3})	a (nm)	η_0 ($\times 10^{-6}$ Pa.s)	E ($\times 10^{-20}$ J)
IW liquid mixture	301	0.3625 [(a)]	1.361 [(b)]	632.8	990 [(b)]	-	-	1.4422 [(b)]	3.007 [(b)]
Latex bead	-	-	-	-	-	1053 [(c)]	80[(b)]	-	-

Table 5.1. *Useful data: (a) Beysens* et al. *(1982); (b) see text; (c) Giddings and Ho (1995)*

5.4. Results and discussion

The procedure to collect data is to vary temperature in logarithmic steps of the difference T - T_c in order to generate a uniform density of measurements when looking at the data in log–log scale. The log–log scale indeed evidences power laws (see Figure 5.3). A total of 43 correlation times have been obtained in the temperature range 0.019–20.6 K. The use of equation [5.12] with laminar flow around the colloid and Stokes damping can be tested. Equation [5.13] gives a velocity $\overline{U} = 3.6 \times 10^{-2}$ m.s^{-1}, using the data of Table 5.1. With liquid mixture viscosity ~ 2.5×10^{-3} Pa.s (Table 5.1), the associated Reynolds number is very low:

$$\mathrm{Re} = \frac{\rho_f \bar{U} a}{\eta} \sim 1\times10^{-3} \qquad [5.18]$$

From the photon-beating correlation time τ_0, an experimental diffusion constant can be obtained by using equation [5.19]. The corresponding values are drawn in Figure 5.3. In the same figure, the data are compared with the calculated Brownian motion diffusion constant using viscosity data, equation [5.12]. Viscosity is deduced from the capillary flow measurements of Allegra *et al.* (1971), whose mean shear rate, as estimated in Oxtoby (1975), leads to ignoring the data at $\varepsilon_c < 1 \times 10^{-4}$, corresponding to the high-shear region (see above section 5.2.2).

Two regions of high and low shear are clearly evidenced in Figure 5.3, depending on the value of $S\tau$ with respect to 1, or equivalently ϵ with respect to $\varepsilon_c = 1.2 \times 10^{-2}$ (see equation 5.3 and numerical values in section 5.2.3; $S = 4.5 \times 10^{-5}$ s^{-1}). Using the total viscosity equation [5.4], in the low-shear region ($\varepsilon > \varepsilon_c = 1.2 \times 10^{-2}$), both diffusion constants take the same values accounting for the fluctuation-induced enhancement of viscosity with respect to temperature. This means that the Brownian motion of the colloids follows the viscosity enhancement, corresponding to an augmentation of the friction forces. The contribution of fluctuations is made more apparent in this low-shear region when we compare the values of the experimental diffusion coefficient with the value inferred from the background viscosity alone (equation [5.5]). In the high-shear region ($\epsilon < 1.2 \times 10^{-2}$), the experimental diffusion constant saturates in relation to the fluctuations deformation, leading to saturation of the viscosity enhancement. In other words, mean field, where the influence of fluctuations is disregarded, becomes relevant, as noted in section 5.2.2.

Fujitani (2021) used a Langevin equation and the dependence of the drag coefficient on the particle speed that accounts for the viscosity inhomogeneities induced by shear to calculate the self-diffusion coefficient. The interesting result is that the diffusion coefficient he obtained compares well with the experimental data (Figure 5.3). The error bars correspond to the maximum deviation due to the approximations made in the shear mapping around the particle.

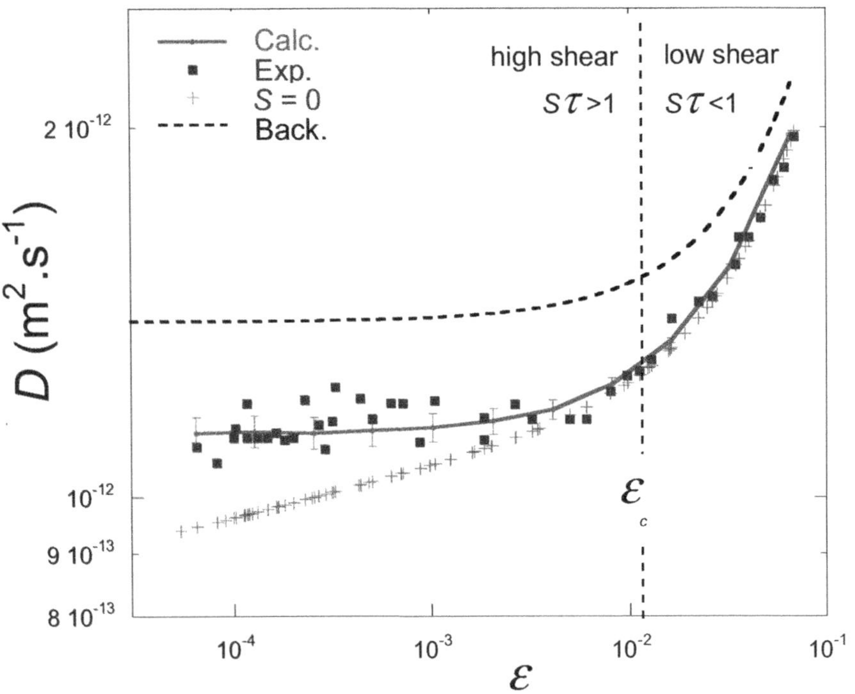

Figure 5.3. *Temperature dependence of the particle self-diffusion constant D (log–log plot) as obtained from the experimental correlation times, equation [5.19] (full squares, exp.) and calculated from equation [5.12] by using full viscosity from Allegra* et al. *(1971) (crosses, S = 0) or only background viscosity from a fit to equation [5.4] and retaining only equation [5.5] (interrupted line, back.). The calculation (Fujitani (2021), full line, calc.) uses the dependence of the drag coefficient on the particle speed and a Langevin equation (see text). For a color version of this figure, see www.iste.co.uk/prudhomme/fluid1.zip*

It is also interesting to compare the D data with other data obtained from PTFE (Teflon) microspheres of 1.15 μm radius immersed in the same isobutyric critical mixture at a critical concentration (Carini *et al.* 1975). The data have been analyzed to determine viscosity using equations [5.19] and [5.20]. Eight data have been obtained in the range T - T_c = 0.6–6.8 K. The transition from high to low shear is expected to occur at T - T_c = 0.06 K, well below the investigated temperature range.

It is worth noting that, in contrast to other colloids immersed in different liquid mixtures (e.g. silica beads in lutidine–water mixtures (Beysens and Esteve 1985; Yabunaka and Onuki 2017)), no colloid aggregation phenomena nor preferential component adsorption were observed on the colloids. The latter would have increased a, i.e. decreased D, an effect in opposition to the saturation effect observed in Figure 5.3.

5.5. Concluding remarks

The Brownian motion of colloids (beads) immersed in a liquid is closely related to the hydrodynamic flows around the colloids. This correlation is highlighted in a near-critical liquid mixture where large concentration fluctuations are present. In such a liquid, the diffusion of immersed colloids is indeed seen to be sensitive to the augmentation of viscosity, related to the critical enhancement of concentration fluctuations. However, as the liquid mixture is brought close enough to the critical point, the fluctuation lifetime turns out to be so large, due to the critical slowing down, that the fluctuations can "feel" the shear. The concentration fluctuations become so distorted by the shear flow that the viscosity enhancement is locally canceled. It results in a saturation of the particle self-diffusion. Such behaviors show that the Brownian motion of particles in a dense fluid can either reflect the fluid properties or modify them, or both.

5.6. Acknowledgements

I gratefully thank S. Perleaux for help with measurements and A. Mongruel for fruitful discussions.

5.7. Appendix: Light scattering (photon beating spectroscopy)

Colloids immersed in a liquid lead to concentration fluctuations and thus to refractive index fluctuations, which scatter light. The dynamics of these fluctuations is driven by the diffusion of colloids and can be analyzed by recording the frequency spectrum of the scattered light or, equivalently, its autocorrelation function (Figure 5.4). In the photon-beating homodyne arrangement used here, the scattered light interferes with itself and the autocorrelation function is a decaying exponential function with typical time τ_0 such that (Cummins 1974)

$$\tau_0 = \frac{1}{2Dq^2} \quad [5.19]$$

Here, D is the colloid self-diffusion coefficient

$$D = \frac{k_B T}{6\pi\eta a} \quad [5.20]$$

and

$$q = 2K_0 \sin\frac{\alpha}{2} \quad [5.21]$$

is the transfer wave vector corresponding to the scattering angle α (Figure 5.4). The incident light wave vector K_0 reads, with n the liquid refractive index and λ_0 the wavelength of incident light:

$$K_0 = \frac{2\pi n}{\lambda_0} \quad [5.22]$$

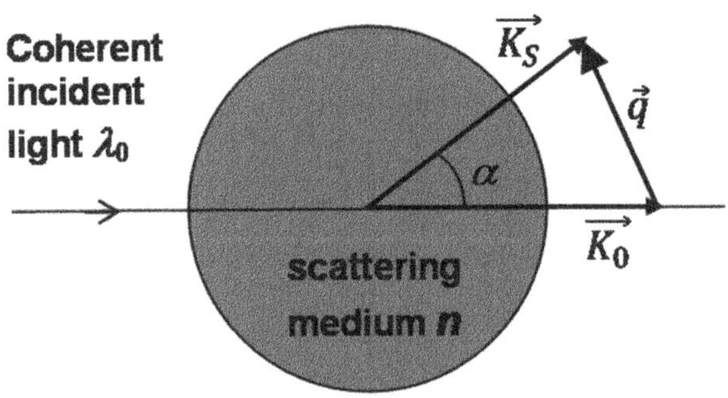

Figure 5.4. *Light scattering arrangement. $\overrightarrow{K_0}$: incident light wave vector; $\overrightarrow{K_S}$: scattered light wave vector; $\vec{q} = \overrightarrow{K_S} - \overrightarrow{K_0}$: transfer wave vector; n: refractive index; λ_0: incident light wavelength. For a color version of this figure, see www.iste.co.uk/prudhomme/fluid1.zip*

5.8. References

Allegra, J.C., Stein, A., Allen, G.F. (1971). Tracer diffusion and shear viscosity for the system isobutyric acid–water near the critical mixing point. *J. Chem. Phys.*, 55, 1716.

Beysens, D. and Esteve, D. (1985). Adsorption phenomena at the surface of silica spheres in a binary liquid mixture. *Phys. Rev. Lett.*, 54, 2123–2126.

Beysens, D. and Gbadamassi, M. (1979). Light-scattering study of a critical mixture with shear flow. *J. de physique lett.*, 40, L565–L567.

Beysens, D., Bourgou, A., Calmettes, P. (1982). Experimental determinations of universal amplitude combinations for binary fluids. I. Statics, *Phys. Rev. A*, 26, 3589–3609.

Carini Jr., G., Maisano, G., Migliardo, P., Wanderlingh, F. (1975). Comparative measurements of viscosity near the critical point. *Phys. Rev. A*, 11, 1755–1762.

Cummins, H. and Pike, E.R. (1974). *Photon Correlation and Light Beating Spectroscopy*. Plenum Press, New York.

Einstein, A. (1905). On the motion of small particles suspended in liquids at rest required by the molecular-kinetic theory of heat [In German]. *Ann. Phys.*, 17, 549–560.

Fujitani, Y. (2021). Suppression of viscosity enhancement around a Brownian particle in a near-critical binary fluid mixture. *J. Fluid Mech.*, 907, A21.

Giddings, J.C. and Ho, J. (1995). Accurate measurement of density of colloidal latex particles by sedimentation field-flow fractionation. *Langmuir*, 11, 2399–2404.

Happal, J. and Brenner, H. (1963). *Low Reynolds Number Hydrodynamics*, 1st edition. Kluwer Academic Publishers, Dordrecht.

Huang, R., Chavez, I., Taute, K.M., Lukić, B., Jeney, S., Raizen, M.G., Florin, E.-L. (2011). Direct observation of the full transition from ballistic to diffusive Brownian motion in a liquid. *Nat. Phys.*, 7, 576–579.

Jacobs, D.T. and Greer, S.C. (1996). Amplitude of the anomaly in the mass density near a liquid-liquid critical point. *Phys. Rev. E*, 54, 5358–5363.

Kheifets, S., Simha, A., Melin, K., Li, T., Raizen, M.G. (2014). Observation of Brownian motion in liquids at short times: Instantaneous velocity and memory loss. *Science*, 343, 1493–1496.

Li, T., Kheifets, S., Medellin, D., Raizen, M.G. (2010). Measurement of the instantaneous velocity of a Brownian particle. *Science*, 328, 1673–1675.

Onuki, A. (2002). *Phase Transition Dynamics*. Cambridge University Press, Cambridge.

Oxtoby, W. (1975). Shear gradient dependence of the viscosity of critical mixtures. *Phys. Lett. A*, 50, 459–460.

Perrin, J. (1909). Brownian motion and molecular reality. *Ann. Chim. Phys.*, 18, 5–104.

Stanley, H.E. (1987). *Introduction to Phase Transitions and Critical Phenomena*. Oxford University Press, Oxford.

Toumi, A. and Bouanz, M. (2008). Volumetric and refractive index properties of isobutyric acid–water binary mixtures at temperatures ranging from 300.15 to 313.15 K. *J. Mol. Liquids*, 139, 55–60.

Yabunaka, S. and Onuki, A. (2017). Critical adsorption profiles around a sphere and a cylinder in a fluid at criticality: Local functional theory. *Phys. Rev. E*, 96, 032127.

Zappoli, B., Beysens, D., Garrabos, Y. (2015). *Heat Transfers and Related Effects in Supercritical Fluids*. Springer, Dordrecht.

Zwanzig, R. and Bixon, M. (1975). Compressibility effects in the hydrodynamic theory of Brownian motion. *J. Fluid Mech.*, 69, 21–25.

6

Basics on Interfaces in Combustion

Combustion results from a set of chemical reactions that release energy. The different chemical bodies involved are, essentially, an oxidant, a reducing agent and the products of combustion. These bodies are in motion and diffuse into each other, and the heat produced by the chemical reactions spreads. There are several sorts of combustive interfaces. We will first approach combustion in a gaseous medium, and investigate both diffusion flames and premixed flames. First we will study the laminar case, and then the action of turbulence on these flames. A few heterogeneous cases are then presented, such as the combustion of plates, sprays and powders.

List of symbols

Symbol	Meaning	Symbol	Meaning
A, B, C, P	Chemical species	s_L^0	Laminar flame speed
B, s	Coefficient of specific reaction rate	t	Time
c_p, c_v	Specific heats	T	Absolute temperature
D	Diffusion coefficient	u	Relative speed
E_a	Activation energy	$\mathbf{V}$	Fluid velocity vector
F	Quantity; fuel	$\mathbf{v}_{//}$	Velocity tangential to the flame
$f(\mathbf{x}, t)$	Position function	$\mathbf{v}_\perp$ $\mathbf{V}\perp$	Velocity normal to the flame

Chapter written by Roger PRUD'HOMME.

f	Quantity F per unit mass	$\mathbf{V}$	Composite velocity in the interfacial layer
g	Thermodynamic potential per unit mass	$\dot{W}$	Chemical production rate
h	Enthalpy per unit mass	x, y, z	Cartesian coordinates
$\mathbf{J}_F\ \mathcal{J}_F, \mathbf{J}_{VF}$	Flux of F for $\mathbf{V} = \mathbf{0}$, $\mathbf{V} = \mathbf{v}$, $\mathbf{V} = \mathbf{V}$	Y_j	Mass fraction of the species j
k	Specific rate of the reaction	j	Index of a chemical species
ℓ_f, ℓ_δ	Diffusion thickness, reaction thickness	β_j, β_T	Reduced concentration, temperature
L	Characteristic length	γ	Isentropic coefficient C_P/C_v
$\mathcal{L}$	Markstein length, phenomenological coefficient	λ	Thermal conductivity
M	Mass, Mach number	μ	Coefficient of viscosity
$\dot{m}$	Unit mass flow rate	ν	Kinematic viscosity $V = \mu/\rho$
n, N	Normal coordinate on small, large scale	Π	Viscous pressure tensor
$\mathbf{N}$	Unit normal	ρ	Density
$p, \mathbf{P}$	Pressure, pressure tensor $(= -\Sigma)$	σ	Surface tension
$\mathbf{q}$	Heat flux vector	Σ	Stress tensor
R	Molar constant of perfect gases	T	Reduced reaction energy
s	Entropy per unit mass	$\vartheta = 1/\rho$	Specific volume
		∇	Gradient operator

6.1. Introduction

The discovery of fire in prehistoric times brought about a true revolution. The energy released by combustion provided heat, making it possible to cook food and extract and work metals (important steps in the stages of human development). It also gave illumination, initially through torches, then candles. Today, combustion has many applications, including in the engines of satellite launchers, all of which run on chemical energy. In this respect, combustion propulsion is currently the only means available in the launch phase, and engine manufacturers seek maximum efficiency and minimal pollution. It must be noted that the cryogenic engines of the Ariane rocket (i.e. engines comprising of a storage stage in a tank at very low temperatures) operate with oxygen and hydrogen, whose combustion product is water and, therefore, non-polluting.

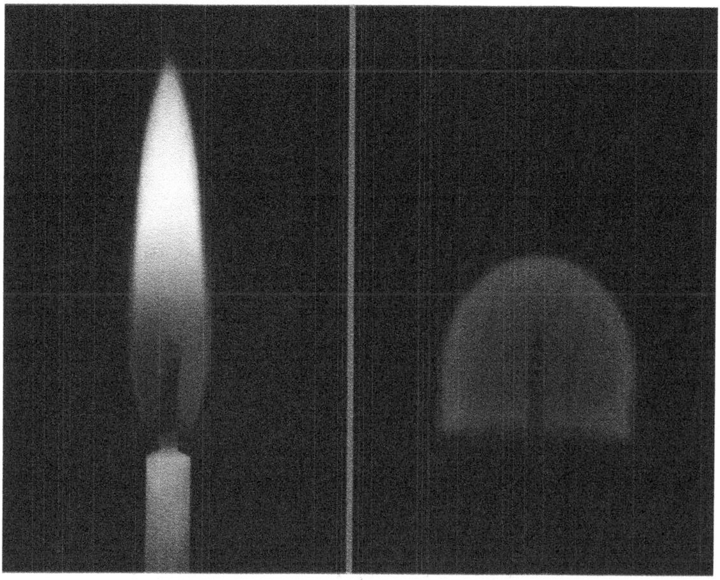

a) b)

Figure 6.1. *Comparison of a flame a) under Earth's gravity and b) in zero gravity on the ISS. Source: https://commons.wikimedia.org/wiki/File:Candlespace.jpg. For a color version of this figure, see www.iste.co.uk/prudhomme/fluid1.zip*

Combustion results from chemical reactions that are generally exothermic, i.e. they produce heat. A classic example is that of the combustion of the H_2–O_2 mixture, which produces water, H_2O, releasing large amounts of heat. In fact, this reaction is made up of a chain of reactions that involve transient, intermediate chemical species, such as the radical[1] OH and isolated H and O atoms. Not every reaction in the chain releases heat, but the chain is exothermic overall.

1 A radical (often called a free radical) is a chemical species that has one or more unpaired electrons in its outer layer, which makes it highly reactive. It is noted by a point.

A chemical reaction can be symbolized simply through A → B, indicating that the reactants, A, yield B, the products of the reaction. Then again, we can write F + O → P, where F is the fuel, O is the oxidant and P encompasses the products of combustion. However, a more detailed description is often required if **chemical kinetics** are to be computed, i.e. if we want to determine reaction rates.

The rate of a reaction is equal to a *specific rate k* multiplied by the products of the reactant concentrations, raised to the power of their stoichiometric coefficient. The specific rate of the reaction is a function of temperature, and is generally expressed in the form:

$$k = \mathrm{B}\,\mathrm{T}^{s} \exp(-\mathrm{E_a}/\mathrm{RT}) \qquad [6.1]$$

where T is the absolute temperature, B and s are constants, R is the universal constant of perfect gases (R = 8.314 J $mole^{-1}$ K^{-1}) and E_a is the activation energy for the reaction.

These rates are heavily dependent on the temperature, as the molecules must receive sufficient energy (called "activation energy") to be able to recombine and produce heat.

Thus, a mixture of hydrogen and oxygen does not react at room temperature as the molecular shocks are generally too weak to break up the H_2 and O_2 molecules, which is required to initiate the reaction. From a certain temperature onwards, the reaction becomes active enough to heat the medium and the rate then increases very rapidly, given that reaction rate is exponentially dependent on temperature. When hydrogen combustion occurs in an entirely gaseous medium, it is said to be combustion in a **homogeneous phase**. The flame from a gas stove or a Bunsen burner is also an example of combustion in a homogeneous phase.

On the other hand, the combustion of a solid in contact with air is said to be combustion in a **heterogeneous phase**. Fires, in general, belong to this category. It must be noted that the concept of homogeneity is very relative, as it depends on the scale of observation. Thus, compressed powders are homogeneous when observed on a scale that is much larger than that of the size of the grains that make up the powder. On the other hand, they appear heterogeneous if observed on the scale of the grains.

Combustion is often a complex phenomenon. Let us take the example of a **candle flame** (Figure 6.1(a)). This brings together the melting of the wax or paraffin, the transport of this melted liquid through the porous wick and its evaporation, followed by its combustion with oxygen in the ambient air. The combustible steam diffuses within the flame, in the products of combustion. Oxygen diffuses to the outside in the mixture of the products of combustion and air. The result is what is called a **non-premixed flame** (or a diffusion flame).

Hot gases tend to rise under the **effect of buoyancy** (the Archimedes principle), given their lower density. This is called natural convection. The flame may oscillate as a result of the development of an instability along the contact surface between the burned gases in motion and external air. This instability, called the **Kelvin–Helmholtz** instability, which also occurs in the case of premixed flames obtained using a Bunsen burner, disappears completely in zero gravity[2].

On the other hand, the presence of carbon oxides (CO), water vapor (H_2O) and soot make the flame bright. This luminosity changes somewhat in *zero gravity*, as natural convection – which elongates the flames – disappears and the flame takes a more hemispherical shape. On the other hand, its color tends towards bluish, as depicted in Figure 6.1(b): in fact, this is a true diffusion flame[3]. The diffusion feeds the flame with oxygen, but the products of combustion, such as CO_2, are not removed through convection and, thus, remain in the vicinity of the flame and tend to block access to oxygen, such that the combustion rate slows down, and the flame is less hot. This explains its color and why it is less bright. Furthermore, the flame does not survive very long in these conditions, lacking the gasification of the fuel.

2 With regard to this, let us note that experiments in microgravity have been very instructional in physics, as they saw the disappearance of the Archimedes thrust, which sometimes changed the aspect of the phenomenon (see also Figure 6.4).

3 http://www.planete-sciences.org/blogs/pla-sci/2014/02/22/le-mystre-de-la-flamme-dans-lespace/.

6.2. Non-premixed laminar combustion[4]

6.2.1. *Overview of a candle flame*

The candle flame is, therefore, part of a complex process that includes a diffusion flame. Figure 6.2 gives a detailed depiction of what happens.

What we call a **flame** here is the contact interface between the gaseous fuel and oxygen in the air: the interior of the flame – which is hollow – is made up of almost pure fuel, while the exterior is oxygen-rich. The diffusion process[5] brings these two components into contact in the reaction zone. This zone thus has a certain *thickness*. This thickness depends on the diffusivity of the gases and also on how temperature-sensitive the reaction is (characterized by the activation energy, E_a, of the chemical reaction).

Efficient diffusive transfers and a high activation energy tend to make the flames thinner, and these then become "hollowed out", in a way. In the case of a candle, the thickness of the flame is millimetric.

The diffusion flame thus develops at the separation between an oxidant-rich zone and another reductant-rich zone, with an optimal concentration ratio for the reaction (the *stoichiometric coefficient*, for monoreactive combustion). The name comes from the fact that the persistence of this type of flame is governed by the diffusion of reactants across the mixture of burned gases on either side of the surface. However, the term **non-premixed flame** is sometimes used instead of "diffusion flame", as the phenomenon of diffusion is not unique to these flames, and is also present in premixed flames, although here it is localized differently, as will be explained in section 6.4. The interior of the flame is essentially heated by radiation from the reaction zone on the surface. This heating, in the absence of oxygen, causes the decomposition of organic vapors in the form of *soot*, chiefly made up of carbon. This white soot (around 1500°C) is what gives the flame its color and thus allows the candle to provide illumination.

4 Let us clarify the title of this section: non-premixed means that the fresh fluids (oxidant and reductant) are separated before coming into contact; laminar means non-turbulent.

5 https://www.encyclopedie-environnement.org/en/physics/diffusion-ultimate-step-good-mixture/.

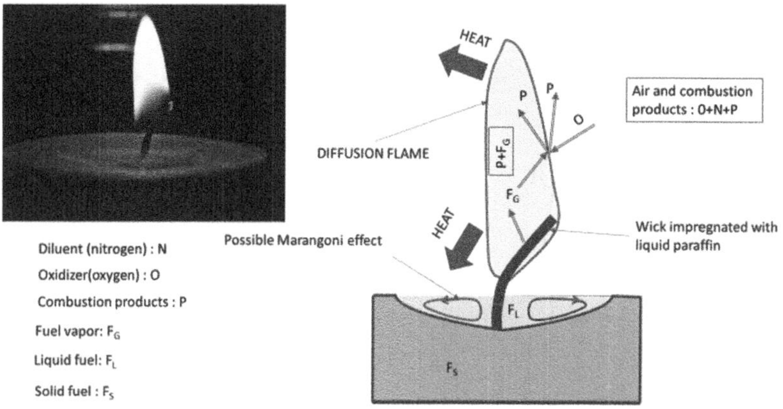

Figure 6.2. *The candle flame, a complex process including a diffusion flame. The diagram shows, in particular, the diffusion of species on either side of the surface of the flame and the heat that causes the paraffin to melt, as well as the agitation of the liquid resulting from the Marangoni effect*[6]*. For a color version of this figure, see www.iste.co.uk/prudhomme/fluid1.zip*

6.2.2. *Analytical solution of a diffusion flame: the Burke–Schumann problem*

It is only the solution to the equations that govern the dynamics of reactive flames that allows us to theoretically predict the structure of the flames. The solution is generally numerical, but there are simplified cases where an analytical computation yields results.

This is what happens in the case of the following example, where the system (which is simpler than the candle) consists of directly injecting the fuel and the oxidant separately. Simplifying hypotheses, such as those of the thin flame in laminar and stationary flow, make it possible to determine the position of the flame and the fields of temperature and of concentrations. The problem may be addressed with two concentric cylinders with circular sections: the fuel F is injected into the inner cylinder and the oxidant O is injected between the two cylinders. This is the **Burke–Schumann problem** (Burke and Schumann 1928).

---

6 The Marangoni effect, named after the Italian scientist Carlo Marangoni, is caused by the non-homogeneous temperature on the heated surface, causing variations in surface tension which, in turn, induce movements in the liquid, resulting from viscosity.

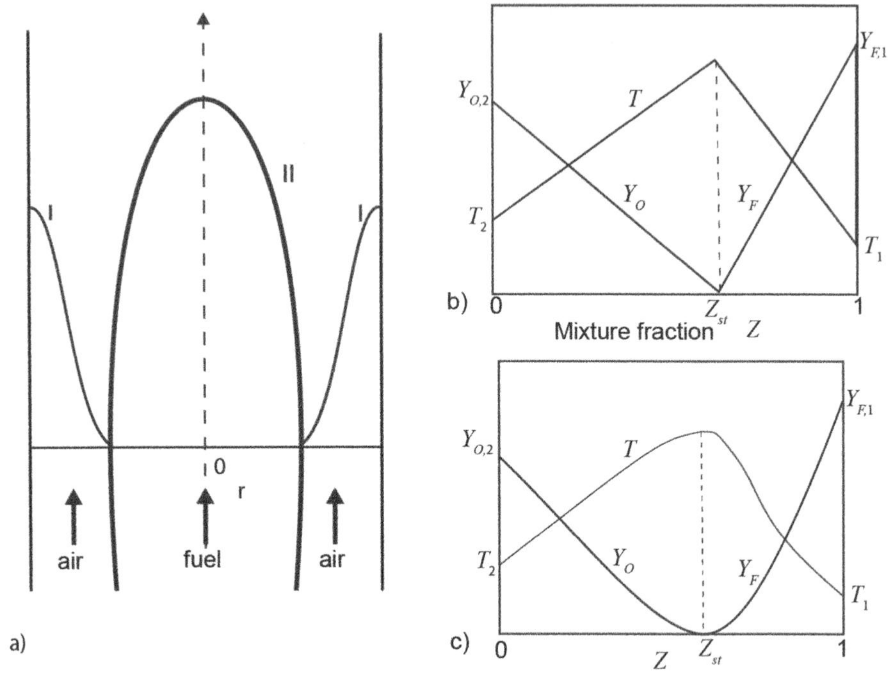

Figure 6.3. *a) Shapes of the Burke–Schumann diffusion flame for coaxial cylinders. Two examples are provided in the same figure: I) under-ventilation, II) over-ventilation; b) solution as a function of Z, corresponding to the theory in the hypothesis where the flame is assumed to be thin; and c) rate of a real solution as a function of Z*

The **laminar flame problem**, entirely governed by the diffusion of the species, can be resolved analytically. The flame, that is assumed to be thin, is positioned at the site of stoichiometric concentration.

The diffusion of species in the axial direction, in front of radial diffusion, is not neglected, and certain simplifying hypotheses are postulated. These are called the *Shvab–Zel'dovich* hypotheses, which are given below:

– stationary flow;

– no thermal diffusion $D_T = 0$;

– negligible external forces at distance;

– negligible viscosity;

– constant, static pressure in the first approximation;

– Fourier's law of thermal conductivity;

– Fick's law for diffusion, and a single diffusion coefficient for all species;

– the Lewis number is close to unity;

– a single reaction;

– mixture of N perfect gases.

The solution is expressed as a *series of Bessel functions*, limited to its first two terms, which explains the somewhat unrealistic shapes of the flame in Figure 6.3(a). However, this makes it possible to clearly differentiate between case I, where the flow rate of the injected oxygen is lower than that of the fuel oil and the flame is, thus, *open* (tulip-shaped) and case II, where the oxygen flow rate is sufficient for the flame to be *closed inside* the inner cylinder.

The position of the thin flame can be identified in the diagram in Figure 6.3(b), which gives the mass fractions of the reactants and the local temperature as a function of the *mixture fraction* Z: the temperature and concentration profiles are rectilinear segments with an angle at the level of the stoichiometric mixture fraction $Z = Z_{st}$.

The mixture fraction Z is:

$$Z = (r_{st}Y_F - Y_O + Y_{O,2})/(r_{st}Y_{F,1} + Y_{O,2}) \qquad [6.2]$$

where $r_{st} = \nu'_O \mathcal{M}_O / \nu'_F \mathcal{M}_F$ is the *stoichiometric ratio* for the chemical reaction:

$$\nu'_F F + \nu'_O O \rightarrow P,$$

with the stoichiometric coefficients ν'_F and ν'_O on the left-hand side, and the respective molar masses of the reactants, $\mathcal{M}_F$, $\mathcal{M}_O$, the indices 1 and 2 corresponding to the supply conduits at F and O, respectively.

Figure 6.3(c) gives the results obtained numerically for a thick flame: the angular profiles are thus smoothened.

6.2.3. *Numerical solutions*

The *numerical calculation* – as shown in Figure 6.4 – makes it possible to determine the more complex profiles of the laminar diffusion flames, appearing when the velocity of the central jet differs from that of the outside jet.

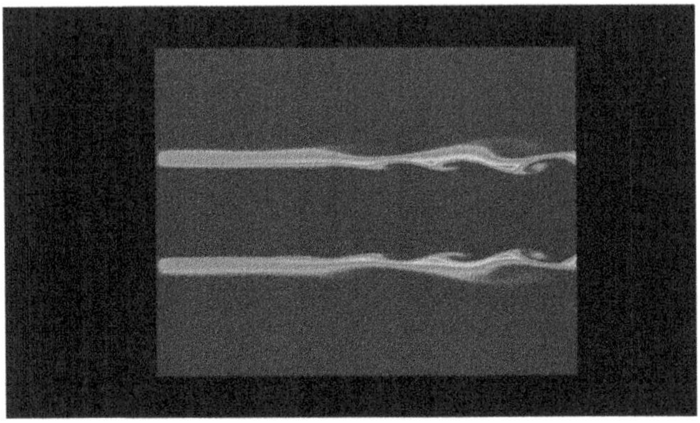

Figure 6.4. *Kelvin–Helmholtz double instability downstream of a running O_2–H_2 injector. Taken from a video created in 2002, kindly provided by the authors C. Nicoli and P. Haldenwang. See also Nicoli* et al. *(2006). For a color version of this figure, see www.iste.co.uk/prudhomme/fluid1.zip*

6.3. Turbulent non-premixed combustion

As René Moreau shows in the *Encyclopedia of the Environment*[7], turbulence accelerates the diffusion of chemical species and of heat, thus promoting combustion chemistry. If the heat diffuses better, the temperature will be higher in the reactive zone, and the chemical reaction will then be that much faster. However, it also has the effect of deforming the flame surface through **wrinkling and stretching**. The many turbulent movements, down to the scale of the smallest vortices, increase the area of contact between the oxidant and fuel. This effectively contributes to increasing the overall rate of combustion.

Figure 6.5 depicts the variety of scales of length that correspond to the sizes of the vortices that can interact with the diffusion flame during the injection of fuel into an oxidant atmosphere.

7 https://www.encyclopedie-environnement.org/en/physics/diffusion-ultimate-step-good-mixture/.

We must *compare the size of the vortices to the thickness of the flame* to characterize the turbulent combustion regime. It is difficult to classify turbulent regimes (see Figure 6.6, which schematizes the Peters diagram (Peters 2000)) in non-premixed combustion, since there is no simple definition for flame thickness.

Scales of turbulence are partly evaluated in the space of the mixture fraction Z.

Figure 6.5. *Visualization of a turbulent jet produced by laser-induced fluorescence. The jet has a wide range of length scales, represented by different colors, which is an important characteristic of turbulent flows. Source: http://en.wikipedia.org/wiki/Turbulence. For a color version of this figure, see www.iste.co.uk/prudhomme/fluid1.zip*

We have:

– A flame thickness $\ell_D = (D_{st}/a)^{1/2}$, where D_{st} is the diffusion coefficient at the position of the flame, when $Z = Z_{st}$ with the stretching rate[8] a.

– The corresponding diffusion thickness in the space of the mixture fractions: $\Delta Z_f = \ell_D \ |\nabla Z|_{st}$. *The diffusion thickness* can be rewritten as a

8 The surface stretching rate is defined in Chapter 1 (section 1.3.2).

function of the stretching rate a and an instantaneous value of the local "scalar dissipation rate" $_{st} = 2D_{st}|\nabla Z|_{st}^{2}$. Thus, we have $\Delta Z_f = (\chi_{st}/2a)^{1/2}$.

– A reaction thickness $\Delta Z_R = \varepsilon\ \Delta Z_f$, equal to the thickness of the oxidation layer, which is proportional to $\chi_{st}^{1/4}$.

Figure 6.6, where Z' is the average fluctuation of the mixture fraction, gives an example of non-premixed, methane–air combustion, with four chemical reactions:

1) $CH4 + 2H + H2O = CO + 4H2O$,

2) $CO + H2O \longrightarrow CO2 + H2$,

3) $H + H + M \longrightarrow H2 + M$,

4) $O2 + 3H2 \longrightarrow 2H + 2H2O$.

Four zones can be seen in the diagram, one of which reveals flamelets.

The concept of a **flamelet**, discovered by Williams (1985a, 1985b), is currently used in all types of turbulent combustion (we must also cite Peters (1980, 2000), Kuznetsov (1982) and Bray (1980; Bray and Peters 1994) for extensions of this concept). Flamelets are thin, reactive–diffusive layers embedded within a non-reactive turbulent flow field. Poinsot and Veynante (2005) give a broader definition of the flamelet regime: a continuous flame front with no quenching, and with no reference to thin flames.

Let us specify that flame quenching occurs when a flame front is subject to external disturbances – such as the loss of heat or aerodynamic stretching – that are strong enough to reduce the reaction rate to a negligible value or, in certain cases, to completely suppress the reaction rate.

We can see that there is no flame in the left side of the Peters diagram, where the scalar dissipation rate is lower than the stoichiometric value, i.e. lower than unity. On the right side, above 1, the flame is lit and it is increasingly deformed as turbulence increases, first through wrinkling and stretching, then as the reaction zone is broken up and the whole flame is finally broken up into separate flamelets with turbulent mixture fractions $Z'_{st}/(\Delta Z)_F$, greater than unity.

The flames produced by injectors in engines (Prud'homme 2015) are the most widely studied, both experimentally and theoretically. Further information on the morphology of these flames can be derived from the diagram in Figure 6.6. This is important when choosing the appropriate turbulent combustion model to numerically analyze flows. We must often consider several regions and a different model must be applied to each.

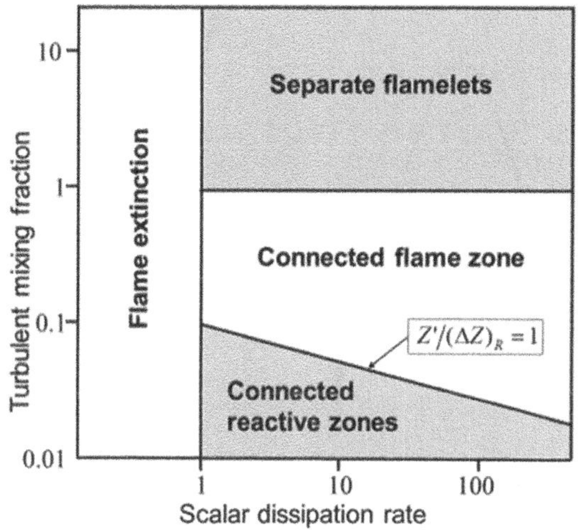

Figure 6.6. *Simplified representation of the Peters diagram (Peters 2000) in the diagram for "dissipation rate $\chi_q/\widetilde{\chi_{st}}$ – turbulent mixture fraction $Z'_{st}/(\Delta Z)_F$" for non-premixed, methane–air combustion regimes*

Models using numerical solutions for combustion are different from each other depending on whether or not the combustion is premixed or non-premixed. Readers who are interested in further information can consult the specialized works we have already cited (Williams 1985a, 1985b; Peters 2000; Poinsot and Veynante 2005).

6.4. Premixed combustion

A Bunsen burner allows us to present laminar premixed flames and to distinguish between these and diffusion flames. As we have seen, combustion is highly restricted by the need to mix the fuel and oxidant. The efficiency of combustion is considerably increased by a premix. We can move from one case to another using a single Bunsen burner, either by positioning its ferrule such that air can enter upstream, thus leading to a premix, or by closing the ferrule, leading to a diffusion flame. In a gas stove, the premix is obtained thanks to the design of the burner, which allows the injection of air. If we light a stove without the burner-top we obtain a larger flame, which is less thermally efficient and produces soot, due to incomplete combustion.

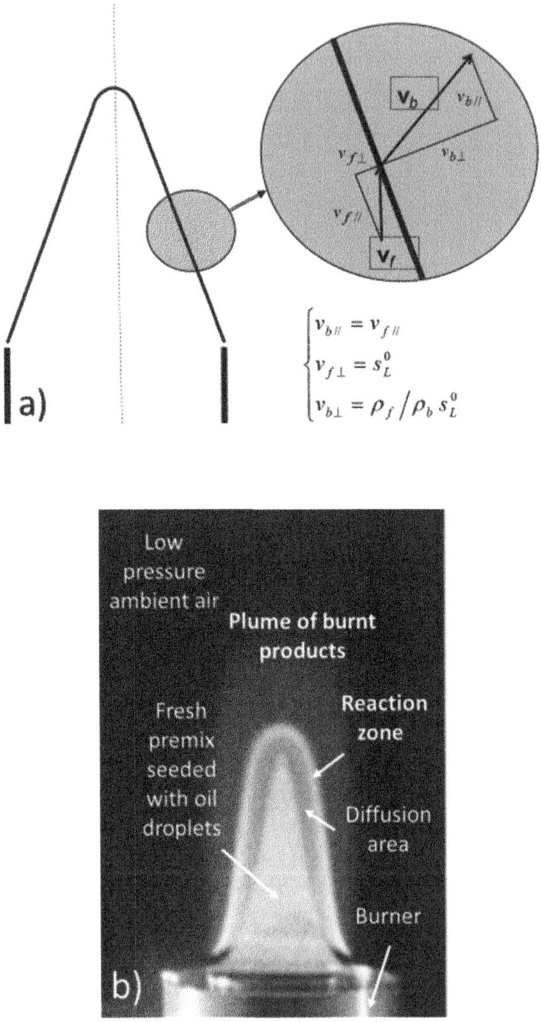

Figure 6.7. *a) Application of the Rankine–Hugoniot theory (see Appendix A, section 6.10.1) to the oblique flame produced by a Bunsen burner: composition of the velocities on either side of the flame front. The burned and fresh gases are denoted by the indices b and f, respectively. b) Laser plan photograph of a Bunsen burner flame under low pressure – which thickens its internal structure – visible due to seeding by fine droplets of oil. Source of b): the Meudon Aerothermal Laboratory. For a color version of this figure, see www.iste.co.uk/prudhomme/fluid1.zip*

It has been seen that a diffusion flame is located at the contact between an oxidant zone and a fuel zone. Thus, it is, in a way, a locus of points at stoichiometric concentration. In a homogeneous premix, on the other hand, the position of the flame cannot be determined *a priori*. The phenomenon which occurs is a wave resulting from interactions that bring into play convection, heat diffusion, and chemical species and reaction, with its activation energy being E_a and the energy released being Δh. In a non-reactive gas, the only discontinuity waves are compression waves, known as shockwaves, which have the property of supersonic speed with respect to the upstream flow. Expansion waves, such as centered waves, do exist, but are continuous.

On the other hand, in a reactive gaseous mixture, two types of discontinuity waves can be produced at the same concentrations. The compression wave exists, as in the non-reactive case. This is a shockwave boosted by the energy of the chemical reaction and is, therefore, more intense and completely supersonic (2000–3000 m/s) and is called a detonation.

However, the laws of thermodynamics make possible the existence of a discontinuous expansion wave – also called a deflagration wave or premixed flame – which is clearly subsonic (of the order of m/s). The shape of the flame produced at the outlet of a Bunsen burner is a result of the equilibrium between the output velocity of the gas and the rate of the propagation of the flame towards the interior (see Figure 6.7). This propagation of a premixed flame is studied theoretically in the simplified case of a rectilinear tube.

6.4.1. *Propagation of a premixed flame*

The simplified Rankine–Hugoniot theory (Barrère and Fabri 1971) (see Appendix A, section 6.10.1) takes into account the different planar, laminar combustion waves that may arise in a **straight, adiabatic pipe** (i.e. with no heat exchange with the outside), initially filled with a mix of oxidant and fuel. Ignition at one end of this tube, using a spark, allows a slow, subsonic wave – called a **deflagration wave** – of the order of m/s, to move through the fresh gases. This is a planar premixed flame. The order of magnitude of the propagation velocity of this velocity is in m/s. After this, the wave accelerates over time and becomes a very fast **detonation wave** that progresses at supersonic speed, of the order of several km/sec.

For this unidimensional system, we have considered an origin linked to the wave, as indicated in Figure 6.8(a), with an upstream zone ($x < 0$) containing fresh gases at constant velocity, separated by a fixed discontinuity surface at $x = 0$ and with a downstream zone containing the burnt gases at constant velocity. The discontinuity equations were written consecutively for mass, momentum and energy, and so were equations of state for the gaseous mixture. The Rankine–Hugoniot relations were obtained after simplification:

$$(p - 1) /(\vartheta - 1) = -\gamma M_1^2 \text{ (flow rate line)} \quad [6.3]$$

and

$$p = [6.(1 - \gamma)\, \vartheta + 1 + \gamma + 2\gamma\, \tau] /[6.(1 + \gamma)\, \vartheta + 1 - \gamma]$$
$$\text{(Hugoniot adiabatic)} \quad [6.4]$$

In this simplified theory, in both cases, the solution points were thus obtained graphically by the intersection of a curve and a line in the volume–pressure plane: deflagration for $M_1 < 1$ and detonation for $M_1 > 1$, with the operating point being found at the intersection of the flow rate line and the Hugoniot adiabatic, which has a hyperbolic shape.

The deflagration wave with velocity $v_1 = s_L^0$ corresponds to a point on the F_1F branch, where the pressure variation is low and where the increase in specific volume indicates heavy dilation. However, it is not possible to find the exact position of this point immediately. The explanation for how this point is determined is given in section 6.4.2.

The point of detonation on the T_1T_2 point is easier to determine.

The detonation wave occurs at the Chapman–Jouguet point, the only stable point on the Hugoniot adiabatic. The evolution of the reduced entropy production $(s_2 - s_1)/c_v$ at the crossing of the wave, as a function of the reduced density ρ_1/ρ_2, indeed presents a minimum at this single point, where the parameters can be calculated analytically. It must be noted that its velocity is determined by the conditions at the point T, where the flow rate line is tangential to the Hugoniot curve. The coordinates of the point T can be easily derived using a simple analytical calculation.

6.4.2. *The combustion rate of the adiabatic, planar premixed flame*

We can determine the parameters of a deflagration wave using analytical calculations in monoreactive cases, and even more easily using numerical calculations. On the other hand, experiments carried out with Bunsen burners yield good results, as the normal velocity along the edges of the cone (see Figure 6.7(a)) is similar to the combustion rate of the planar flame (neglecting the effects of curvature). The photo in Figure 6.7(b), taken under a pressure of 1000 Pa, shows the same three zones as the planar flame in Figure 6.8(b) – observed on a small scale in a straight adiabatic pipe.

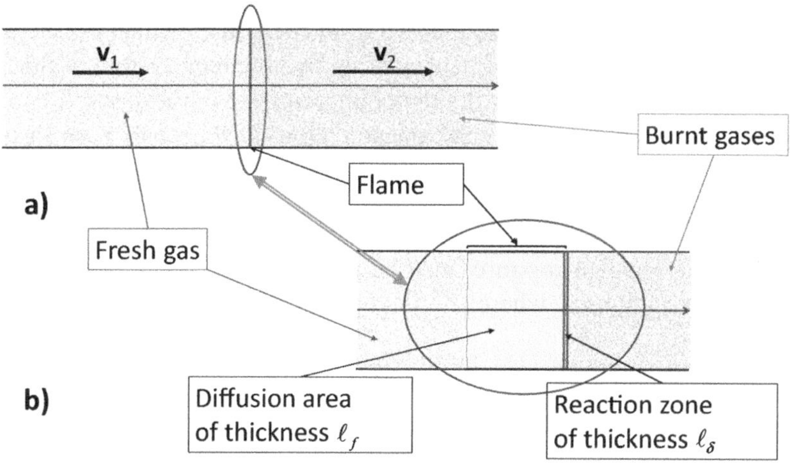

Figure 6.8. *Two scales for the same problem. a) The Rankine–Hugoniot configuration: representation of the stationary motion on the laboratory scale in a system associated with the wave; straight adiabatic pipe initially filled with a mixture of oxidant and fuel, v_1 is the adiabatic laminar combustion rate, generally denoted by s_L^0. b) Structure of the thin premixed laminar flame in the rectilinear conduit on a small scale: the flame itself, with thickness ℓ_f, presents a diffusion–convection zone and a reaction–diffusion zone with thickness ℓ_δ. The activation energy of the reaction here is very high and satisfies $\ell_\delta << \ell_f$. For a color version of this figure, see www.iste.co.uk/prudhomme/fluid1.zip*

However, as Williams (1985a, 1985b) showed, it is possible to evaluate the combustion rate s_L^0 through reasoning – and elementary calculations – based on the fundamental qualities that govern the physics of a deflagration wave.

In order to do this, we must review the list of fundamental quantities that come into play in this process. There are, consecutively:

– the thickness of the flame, similar to the thickness of the diffusion zone ℓ_f in Figure 6.8(b);

– the velocity of the fresh mixture $v_1 = u_1 = s_L^0$;

– the heat of the reaction Δh (energy released per unit mass of the reactant mixture);

– the rate of the reaction $\dot{W}$ (mass of the mixture of reactants converted into the product of the reaction per unit volume per second).

It is possible to evaluate the thermal gradient in the diffusion zone using the relation $dT/dx \sim (T_2 - T_1) / \ell_f$, considering that the temperature of the fresh gases, T_1, is uniform, and that the temperature of the burnt gases, T_2, is also uniform, as the walls of the tube do not allow any heat to pass through. Further, by the law of conservation of energy, we have: $\Delta h = c_p (T_2 - T_1)$.

In order for the process to be stationary, the energy released per unit volume of the reactant mixture must be equal to the energy transmitted by conduction per unit area, where: $\Delta h \, \dot{W} \, \ell_f \approx \lambda \, (T_2 - T_1)/\ell_f$.

The stationary status also implies the conservation of flow; therefore: $\rho_1 \, u_1 = \dot{W} \, \ell_f$.

Several interesting relations can be derived. First of all, the flame thickness can be evaluated as

$$\ell_f \approx (\lambda / cp \, \dot{W})^{1/2} \qquad [6.5]$$

The **combustion rate** $u_1 = s_L^0$ is written:

$$u_1 \approx (1/\rho_1) \, (\dot{W} / cp)^{1/2} \qquad [6.6]$$

However, the above relations lead to important physical results.

First of all, it is possible to compare the flame thickness with the mean free path of the gaseous molecules, as well as with the laboratory scale.

Indeed, it is possible to measure the velocities of the flame[9]. Using representative values for λ, c_p and ρ_1, and with $u_1 \approx 50$ cm/s, the aforementioned equations give $\ell_f \sim 10^{-2}$cm. Consequently, ℓ_f is large when compared to a mean free path of a molecule of about 10^{-5}cm, and the equations for the fluid dynamic continuum are valid in the deflagration wave. However, the same ℓ_f is small when compared to the typical dimensions of the experimental equipment (e.g. the diameter of the mouth of the burner – and therefore the curvature radius of the flame cone – for experiments with Bunsen burners), and laminar deflagration waves may be treated as discontinuities in many experiments.

On the other hand, little is known about the reaction rates and, in practice, the values of $\dot{W}$ are so uncertain that the relation for the combustion rate is more useful in estimating $\dot{W}$, starting from the experimental values of u_1, than the other way around.

The characteristics of the combustion reactions can also be better deduced from the characteristics of the flames. A useful way of deriving an order and the overall activation energy of a reaction consists of measuring the dependencies of u_1 on T and p, and empirically correlating these with the following equation, by adjusting n and E_1.

Indeed, if $\dot{W} \approx p^n e^{-E_a/RT_1}$ is accepted as the reaction rate, where n is the order of the chemical reaction[10], i.e. the number of reactant moles coming into play on the left-hand side, we find:

$$u_1 \approx p^{(n/2)-1} e^{-E_a/2RT_1} \qquad [6.7]$$

These simple calculations are useful especially in evaluating the influence of the temperature of the burned gases $T_1 = T_b$, on the thickness ℓ_f of the flame and on the combustion rate. The pressure p, which varies very little across the flame, also comes into play.

But of course, only a detailed study of the structure of the planar deflagration wave, schematized in Figure 6.8(b), where the abscissa x has

9 For premixes with air, the velocities measured were, for example, 0.845 m/sec with methane and 1.845 m/sec with ethylene (values from the *Encyclopedia Universalis*).

10 For example, n = 2 for A + B → P or n = ν for νA → P.

been artificially dilated, makes it possible to precisely determine its **adiabatic velocity** s_L^0. Direct analytical calculation is possible in simple cases. For this calculation, we use the Shvab–Zel'dovich approximation, introduced in section 6.2.2 for diffusion flames, but which can also be used in the case of premixed flames[11]. The *analytical calculation* is conventional and is found in most basic texts on combustion (Prud'homme 2010). This method will not be presented in detail here, and we will restrict ourselves to presenting the main results.

The calculation for this continuous evolution gives

$$s_L^0 = (\mathrm{M}\, p/\rho \mathrm{R})\, (\rho \mathrm{D\, B\, T^s}/\Lambda)^{1/2} \qquad [6.8]$$

with Λ an eigenvalue that ensures the continuity of the evolution through the reaction zone, B and s the chemical kinetic constants, defined in section 6.1, and M the mean molar mass of the mixture.

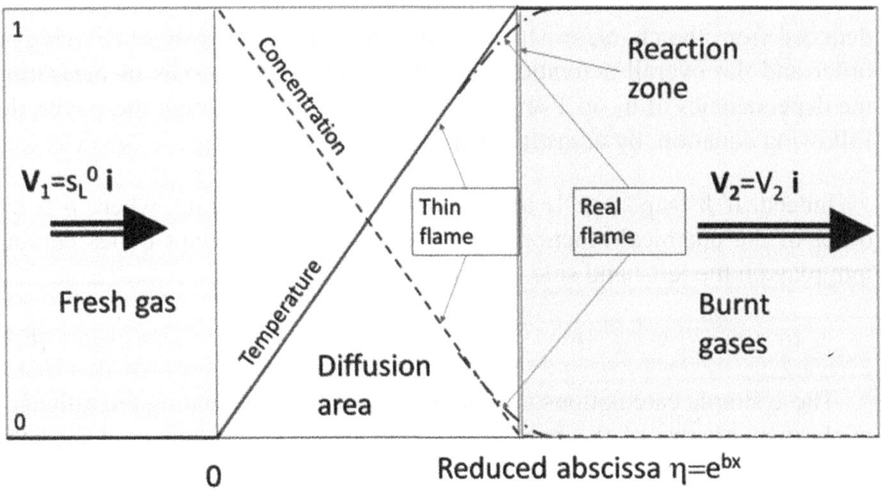

Figure 6.9. *Evolution of the reduced temperature and concentration in a deflagration wave, as a function of the space variable η, with b = v_1/D being constant. The solid lines and dotted lines show the results of the analytical calculation. The dots-and-dashes line shows the passage through a reaction zone for a real flame. For a color version of this figure, see www.iste.co.uk/prudhomme/fluid1.zip*

11 When the Lewis number (Le = $\lambda/\rho \mathrm{D}\, c_p$) is close to unity, which means that the diffusion of species and heat diffusion – i.e. thermal conductivity – take place at the same velocity.

Figure 6.9 shows the evolution of reduced temperature and concentration as a function of the space variable (here, the reduced abscissa). The reduced temperature is $\beta_T = (T - T_1) / (T_2 - T_1)$. For a single reaction $A + B \rightarrow P$, the concentration variable is $\beta = Y_A / M_A = Y_B / M_B$.

The evolutions are continuous in the diffusion zone and in the passage through the reaction zone, where there is a discontinuity of the slope. This slope discontinuity gives angular points and is thus replaced by a continuous evolution of the slope in the case of real flames (dots-and-dashes line), where certain restrictive hypotheses are no longer necessary.

More modern analytical methods lead to the application of the *connected asymptotic expansion* method on three scales: the scale of the flow ℓ, that of the diffusion scale ℓ_f and that of the chemical reaction ℓ_δ (Peters 2000).

It is essential that we use *numerical calculation* for multi-reaction systems and where the simplifying hypotheses are not valid.

6.4.3. *Concepts in turbulent premixed combustion*

The action that turbulence has on premixed flames has been studied and modeled, as with diffusion flames. The representation of turbulent combustion regimes takes place in the turbulence intensity plane as a function of the scale of turbulence. Unlike non-premixed flames, the spatial scales of combustion can be directly compared with those of turbulence (Peters 2000; Poinsot and Veynante 2005).

It must be noted that it was with respect to premixed flames that Damköhler (1940) first presented theoretical expressions for the rate of turbulent combustion s_T.

Starting from the equivalence between the mass flow of a laminar flow across the flame surface with area A_T and the mass flow of the turbulent flow through the cross-section A, he derived the following relation between the laminar and turbulent combustion rates of a stationary flame in a tube: $\dot{m} = \rho_f s_L A_T = \bar{\rho_f} s_T A$, where the overline indicates the average (Figure 6.10).

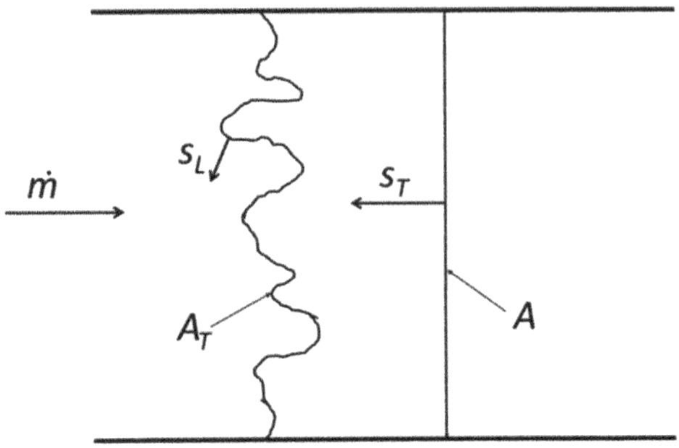

Figure 6.10. *Effect of turbulence on a thin, planar premixed flame: the Damköhler relation* $\dot{m} = \rho_f s_L A_T = \bar{\rho_f} s_T A$

6.5. Plate combustion

A classic example of heterogeneous combustion is that of the combustion of plates, which has been of great interest in studies into preventing fires. This interest has led to many experiments both on the ground and in zero gravity, where one of the main issues is the emission of soot, which can be highly damaging within a spacecraft.

Figure 6.11 gives examples of flames obtained using experimental simulation devices. The flames obtained by the combustion of PMMA, for example (Herbert *et al.* 2013), are not easy to use with different measuring techniques (due to the propagation of the flame, the time taken to change a plate, etc.). Because of this, gas burners have been developed to produce a stationary flame that can be ignited and extinguished quickly. The gas burner in Figure 6.11(a) is made up of a porous bronze material, with a thickness of 20 mm and, when vertically positioned, its size is 400 mm x 250 mm. It is fixed before a stilling chamber from which the gas emerges horizontally. Those in Figure 6.11(b) and (c) are made up of a square porous part arranged at the beginning of the plate in the upstream part, through which the gaseous fuel is vertically injected at velocity V_H and the oxidant circulates, in parallel, at velocity V_O.

A simplified modeling of the problem of the combustion of an ablative plate has been provided by Emmons (1956): the fuel plate, placed in a laminar airflow with a low Reynolds number, produces a diffusion flame upon ignition. The heat causes the surface of the solid to melt, and the resulting liquid evaporates. The fluid flow is compared to the flow of the Blasius boundary layer, a classic flow in fluid mechanics, above a flat plate. This analytical modeling of the combustion problem leads to a thin, stationary laminar flame profile with a parabolic shape. It is normal for this result not to be satisfied during experiments on the ground, given the simplifying hypotheses used, including the hypothesis where gravity is neglected. However, the experiment conducted in microgravity shows a configuration case with an Emmons boundary layer quite well (Figure 6.11(c)).

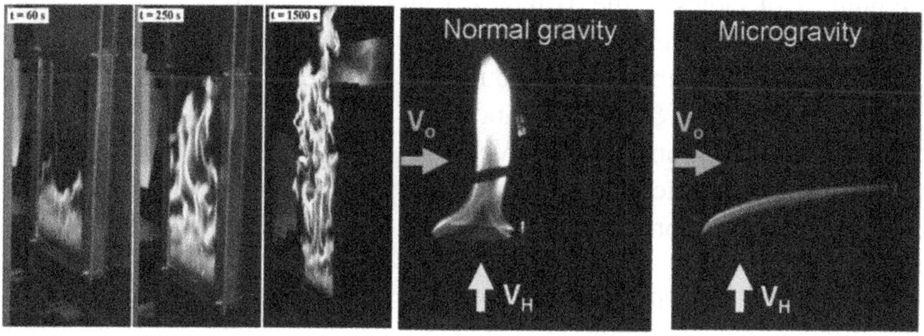

a) b) c)

Figure 6.11. *Combustion of plates through experimental simulation devices: a) evolution photos taken during the combustion of a vertical PMMA plate. Permission given by Herbert* et al. *(2013); b) and c) show experimental observations of a diffusion flame above a plate: b) under normal gravity and c) in microgravity (from G. Legros, private communication in 2009 (Prud'homme 2015)). For a color version of this figure, see www.iste.co.uk/prudhomme/fluid1.zip*

For this two-dimensional stationary calculation, we must write the equations for total mass, momentum, the mass of the species and energy. The form adopted shows the reaction rate on the right-hand side of the last two relations. Discontinuity balance equations are established at the level of the plate. The discontinuity of the mass balance equation reveals the normal ablation flow, ρ_s v_a, which is the same as the combustion rate and the effort exerted on the plate, τ_{xy}. The ablation rate is written as:

$$v_a = - (U_\infty / 2\xi)^{1/2} f(0) (\rho\mu)_P / \rho_H \qquad [6.9]$$

In this expression, the function $f(\xi)$ satisfies the Blasius equation:

$$f''' + f f'' = 0 \qquad [6.10]$$

with two conditions with conventional limits – $f'(\infty) = 1$ and $f'(0) = 0$ – and one condition with a new limit:

$$f(0) = -B\, f''(0) \qquad [6.11]$$

where $f = f(\xi)$, and where the Spalding parameter B comes into play, $B = (\beta_{T\infty} - \beta_{TP} - \beta_{O\infty})\, \Delta h/l$, Δh and l being, respectively, the energy of the chemical reaction and the phase change heat (in the conventional, non-reactive case $B = 0$). On the other hand, as the chemical reaction is $H + O \rightarrow P$, $\beta_O = n_O$ is the number of moles of the oxidant per unit mass and $\beta_T = c_p\,(T - T^0)/\Delta h$ is the reduced sensible enthalpy, with T^0 the reference temperature. Finally, as the fluid is compressible, we introduce the Howarth transformation to define the space variables $\xi = \int \rho\mu dx$ and $\zeta = \int \rho\, dy$. The quantity $-f(0)$ is a well-defined function of B. We can easily obtain the changes in concentration and temperature in the plane. Figure 6.12 shows the configuration of the problem and the curve $-f(0)$ as a function of B.

In order to be applicable, the Emmons theory requires the use of the Shvab–Zel'dovich hypotheses. Several factors could lead to errors, including:

– gravity, which causes hot gases to rise (Figure 6.11(a) and (b));

– the presence of soot, which leads to radiation losses;

– the loss of enthalpy along the thermo-conductive wall and during pyrolysis.

It has been proposed that the Spalding B parameter be modified and replaced by the Torero B_T parameter, which takes into account these last two factors, but not the Archimedes principle (Prud'homme 2010). The use of this new parameter has proven successful in microgravity experiments.

Soot is formed during the combustion of carbon species. The nucleation and increase in soot, as well as its interaction with reactive flow, have been studied by many authors. Taking the soot into account required the development of appropriate mechanisms and taking its interaction with the flow into account. Dorey (2012) carried out a comprehensive review of this complex problem. There are four distinct processes: 1) nucleation, 2) growth on the surface, 3) oxidation and 4) coagulation. Then we must consider the interaction of the soot formation mechanisms with the combustion flow and radiation transfer.

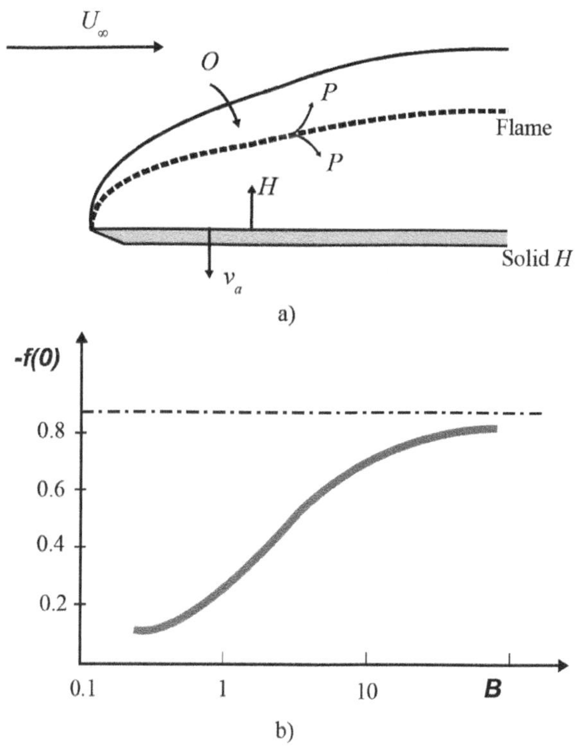

Figure 6.12. *a) The configuration of the Emmons problem and b) curve (-f(0)) as a function of B. For a color version of this figure, see www.iste.co.uk/prudhomme/fluid1.zip*

6.6. Powders

6.6.1. *Thermites*

An example of a **solid medium** that produces a highly exothermic reactive phenomenon **in the absence of gas** is that of **thermites**, powdery mixtures of metallic aluminum and another metal oxide.

Figure 6.13. *Using thermite for soldering rails. The temperatures attained during chemical reactions in the solid phase between aluminum and iron oxides are high enough to allow welding. (source: https://commons.wikimedia.org/wiki/File:Thermite_welding_16.JPG). For a color version of this figure, see www.iste.co.uk/prudhomme/fluid1.zip*

These have long been used, for example, to weld railroad tracks (Figure 6.13). The temperatures attained during chemical reactions of aluminum in the solid phase with iron oxides are high enough to allow welding, as seen in the following examples: Fe_3O_4 and Al (3088°C), FeO and Al (2500°C), Fe_2O_3 and 2Al (2960°C).

6.6.2. *Solid propellant rockets*

A derivative example is that of **solid propellant rockets**[12] (Figure 6.14). The powder cake fitted into the body of the rocket is hollow. When ignited, it releases gaseous species and liquid gas (aluminum, for powders that contain this metal (Glorian 2014)) under the effect of heat. Combustion is

12 https://en.wikipedia.org/wiki/Solid-propellant_rocket.

manifested in the **heterogeneous fluid** jet. The temperatures reached are very high, as is the resulting thrust.

One of the problems encountered is the accumulation of solid alumina Al_2O_3 – a product of the reaction – within the engine, upstream of the outlet pipe. Research into the combustion modes of powder is still ongoing.

Let us take the case of the first stage of the euro-Italian Vega rocket, built using Ariane 5 technology. This functions on **butalane** 68-20 associating 12% of the polybutadiene-based binder (buta), 68% of ammonium perchlorate NH_4ClO_4 (l) as oxidant, with 20% aluminum (ane) as fuel.

The combustion process for butadiene is complex and involves reactions in the gaseous phase and surface kinetics. The typical flame temperature for an aluminized propellant (like Butalane®) is 3500 K (2300 K without aluminum). The combustion of particles thus brings in a considerable amount of energy. The aluminum burns a few mm away from the solid wall in the gaseous phase, and causes very little change in the combustion of the propellant (the flame is located about 100 μm from the surface). At this temperature, we once again have the final stable products of the decomposition of the propellant in the gaseous phase, namely: N_2, H_2O, CO, CO_2, HCl. The micrometric particles, whose surface is covered with alumina (the thickness of the alumina layer is around 50 Å), detach from the propellant surface and partially agglomerate close to the surface, until they reach diameters of around 100 μm. 70% of these are found right in the flow from the propulsion chamber. In a second phase, the agglomerates burn, releasing residues formed by small particles of alumina (about 1 μm).

The aluminum agglomerates usually combust completely as long as they remain within the engine for sufficient time. This complete combustion of the agglomerates leaves alumina residues whose size is of the order of 10 μm.

The schema in Figure 6.15 is valid for the combustion of every drop of liquid aluminum.

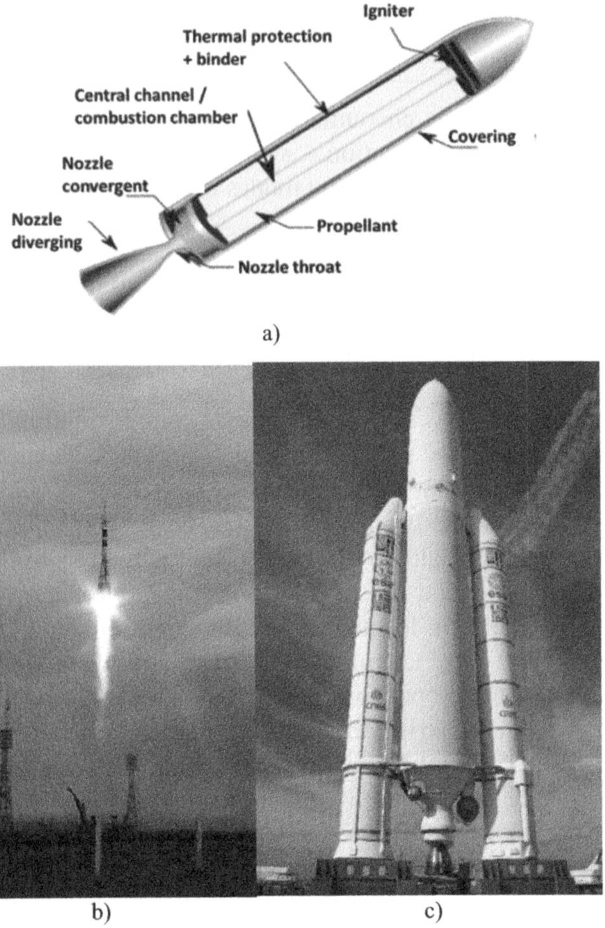

Figure 6.14. *a) Schematic of a solid propellant rocket (source: https://fr.wikipedia.org/wiki/Fichier:SolidRocketMotor-fr.svg). b) Launch of the Soyuz rocket carrying cosmonauts to the ISS on October 21, 2001, from Baïkhonour[13] c) The European Ariane rocket (source: https://fr.wikipedia.org/wiki/Fichier:Ariane_5_Le_Bourget_FRA_001.jpg[14]). For a color version of this figure, see www.iste.co.uk/prudhomme/fluid1.zip*

13 Since the American Space Shuttle was retired in July 2011, the Soyouz shuttle has been the only vehicle capable of sending humans into space (along with Chinese rockets). The various stages are equipped with liquid-propellant engines. Since May 27, 2020, the SpaceX company has provided the USA with a new launcher and capsule.

14 The Ariane 5 launchers have a lower part that contains a central body equipped with a Vulcain cryogenic engine, and also two powder-propelled acceleration stages used to supply the essential thrust during take-off.

The above description excludes any over-simplified reactive model. If, for example, we study the formation of alumina, we must take into account not only the gaseous phase reactions, but also the adsorption–desorption and chemical sorption that bring into play many other intermediate species present at high temperatures, as can be seen in Figure 6.15.

The single heterogeneous mechanism can include 57 elementary reactions between 58 species, of which two are with surface species.

The formation of Al_2O_3 in the presence of air, with the index $_L$ meaning "liquid", involves the following five reactions:

$$2Al + 3/2\ O_2 \rightarrow Al_2O_{3L};\ 2AlO + 1/2\ O_2 \rightarrow Al_2O_{3L};$$

$$AlO_2 + AlO \rightarrow Al_2O3_L;\ Al_2O + O_2 \rightarrow Al_2O_{3L};$$

$$Al_2O_2 + 1/2O_2 \rightarrow Al_2O_{3L}.$$

If we consider the species generated by the propellant, there are also:

$$Al_2O + 2\ CO_2 = Al_2O_3 + 2\ CO;\ 2AlO + CO_2 = Al_2O_3 + CO;$$

$$Al_2O + 2H_2O = Al_2O_3 + 2H_2;\ 2AlO + H_2O = Al_2O_3 + H_2.$$

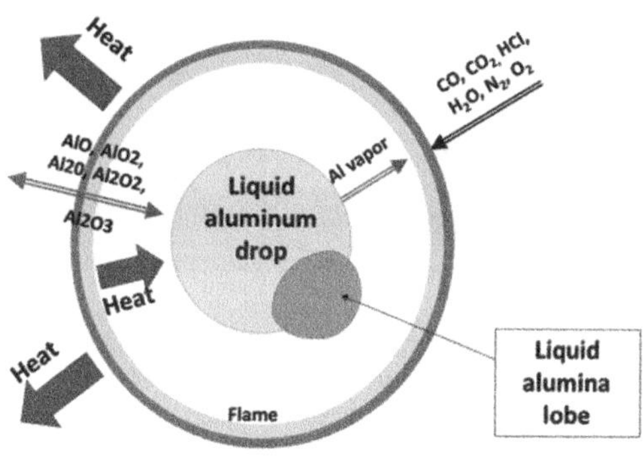

Figure 6.15. *Schema for the combustion of an aluminum particle in a propellant environment. With the permission of the author, Glorian (2014). For a color version of this figure, see www.iste.co.uk/prudhomme/fluid1.zip*

Only a numerical calculation can solve a problem like this, which also contains information relative to the various means of transport, such as convection, diffusion and radiation.

6.6.3. *Spin-like combustion*

Let us take a final example of the **combustion of solids**, which may prove to be very useful when collecting chemical kinetic data.

As indicated in section 6.4.2 for premixed flames, measurements carried out in the combustion of solids are also useful in determining quantities in chemical kinetics. Holt *et al.* (1985) were able to use, for example, the combustion of TiB_2. Other synthetic reactions have also been studied for products that are useful for electronics and battery packs. Here, we thought it would be useful to present a mode of surface propagation for these combustions that has been the subject of recent research. Studies have been carried out on propagation in a spiral of combustion nucleation sites. This is the surface combustion of metallic powder cylinders. Kirdyashkin *et al.* (2007), in particular, experimentally studied powder combustion waves under a laser beam (Klimenok and Rashkovskiy 2015).

Figure 6.16. *Propagation of a discrete combustion wave in multilayer spin mode (the light-colored particles are burning; the dark particles are non-flame retardant). The combustion takes place towards the bottom; the spin moves along the surface, from right to left. Extract from the article by Klimenok and Rashkovskiy (2015) with the publisher's permission. For a color version of this figure, see www.iste.co.uk/prudhomme/fluid1.zip*

More recently, Klimenok and Rashkovskiy (2015) published a discrete model for the "spin combustion" of powder mixtures in the absence of gas. Bharath *et al.* (2013) believe that the process may sometimes be purely thermal in nature and related to the heterogeneity of the dispersed systems and to the existence of a temperature threshold. Figure 6.16 shows the progression of the surface combustion, notably of the propagation of a discrete combustion wave in multilayer spin mode.

6.7. Sprays and fireworks

A heterogeneous character is also seen with liquid propellants. This is most often seen in spray combustion (Figure 6.17).

Knowledge of the size distribution of the sprays is essential for obtaining optimal combustion in the engines and, thus, producing maximum energy while minimizing the quantity of pollutants produced[15].

This holds true for land-based engines as well as airplane and rocket engines. In the case of the cryogenic engines of the Ariane rocket, the spray is made up of drops of liquid oxygen within the gaseous hydrogen and, furthermore, the operating conditions are supercritical[16].

Combustion instabilities may arise. This is a crucial problem as these instabilities may cause significant damage, or even the destruction of the rocket. Certain instabilities are caused by the amplification of vibrations from the engine, resulting from the effect of the evaporation of the droplets from the injectors.

Another heterogeneous situation is manifested in hand-held **fireworks**. Inoue *et al.* (2017) studied the cascading fragmentation problem. To this end, they examined, in particular, the evolution of a droplet disturbed by the explosion of an internal gaseous cavity. They thus demonstrated that the drop may oscillate around its initial spheroidal shape when the energy involved is smaller than a certain critical value. When the energy involved is high enough, the drop may deform into a ligament that is elongated enough to get fragmented.

15 https://www.encyclopedie-environnement.org/en/air-en/air-pollution-particles-what-are-they/.

16 Above the critical liquid–vapor point, there is no longer any difference between a liquid and a vapor; there is only a single fluid called a supercritical fluid. The interface between liquid oxygen and gaseous hydrogen no longer presents surface tension, which was present in the subcritical regime. The critical point is a singular point. In its vicinity, the thermal conduction coefficient tends to zero. A new mode of heat transfer appears, called the piston effect.

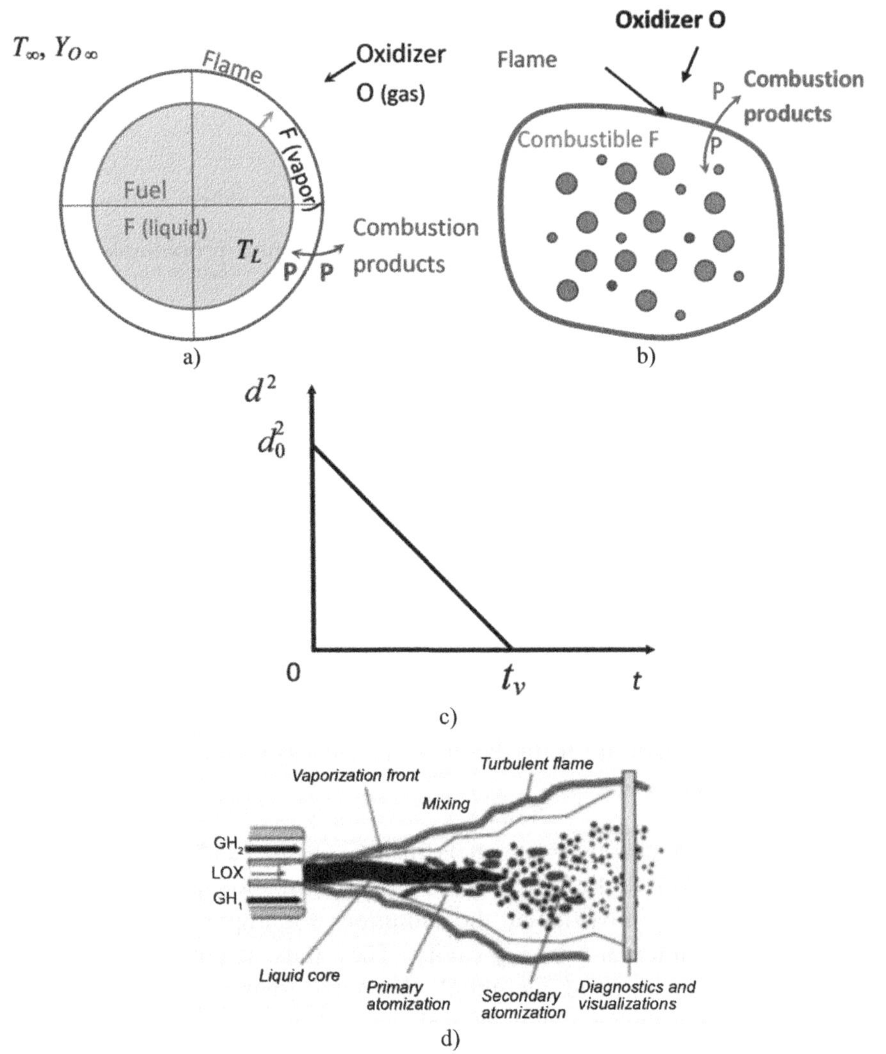

Figure 6.17. *Spray combustion: a) individual drop combusting with a spherical diffusion flame*[17]*; b) packet of combusting droplets, each drop evaporating; c) in both cases, the evolution of the diameter of the drop,* d*, closely follows the law:* $d^2 = d_0^2(1 - t/t_v)$*,* t_v *lifespan; and* d) *the physical phenomenon induced in the flame from a cryogenic coaxial injector (according to the ONERA website). For a color version of this figure, see www.iste.co.uk/prudhomme/fluid1.zip*

---

17 Spray combustion is taken as an example of a heterogeneous flow. However, at the scale of the drop, it is in fact a diffusion flame in a homogeneous medium.

Figure 6.18 demonstrates the combustion of a sparkler. We can clearly see the traces of the droplet paths illustrating the continuous cascading fragmentation process in a traditional firework.

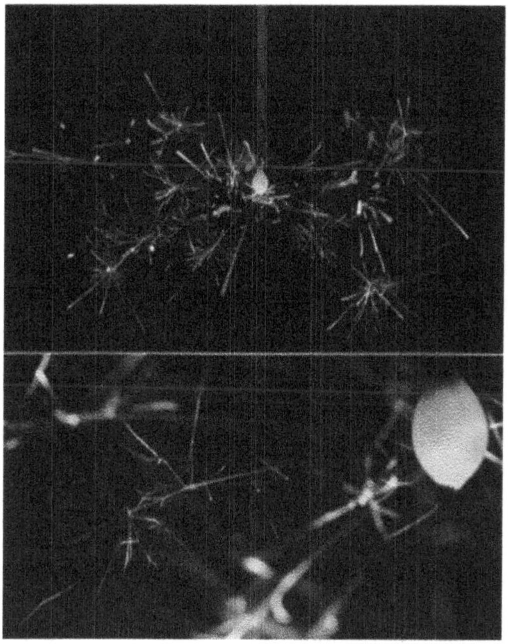

Figure 6.18. *Traces of the droplet paths illustrating the continuous cascading fragmentation process in a traditional firework. On top: exposure time 1/60 s; the paper cord with a 2 mm diameter retains the mother drop. Bottom: an exposure time longer than 1/25 s reveals the cascade of consecutive ruptures, with up to four generations visible on the in-focus arborescence. With the permission of the publisher (Inoue* et al. *2017). For a color version of this figure, see www.iste.co.uk/prudhomme/fluid1.zip*

Like the fireworks, the flames emit light as they contain bodies at high temperatures. We will not elaborate on the radiation mechanism behind this, as this chapter focuses on combustion interfaces.

6.8. Conclusion

– Combustion interfaces are premixed or non-premixed flames, evaporation surfaces, thermites, cavitation pockets, solid or liquid boundaries.

– Combustion results from the release of energy from the "reactants" through chemical reactions that yield "products of reactions", said to be "burnt products of reaction".

– This may occur in a solid or fluid medium that may be homogeneous or heterogeneous. In gaseous mixtures, combustion leads to the formation of flames of different kinds, which develop in laminar or turbulent flow. The solids are generally compacted powders. Radiation often plays an active role.

– It is studied for research or forecasting purposes and requires balancing physical quantities that come into play in these mixtures: the mass of species, momentum, energies. This implies a knowledge of chemical kinetics and the constitutive laws of the medium. The solution, generally a numerical solution, has given rise to the development of many specific methods that are beyond the scope of this discussion.

This is a very vast subject that plays a role in multiple domains and which has a rich history. Some current research includes, notably, nanoparticles generated by flames (used in medical technologies or used as essential additives for computer chips), the reduction of pollutants (like soot or aromatic emissions from buses and airplanes), the spread of flames (to prevent and control urban fires and forest fires), electricity generation (from clean, carbon-free fuels) and so on.

For updated and further information on combustion, it would be useful to consult the following journals: *Combustion and Flame, Combustion Science and Technology*, and the *Proceedings of the Combustion Institute*. For more on "spin combustion" in powders, see Maksimov *et al.* (1978), Aldushin *et al.* (1981), Rashkovskiy *et al.* (2010) and *Combustion Explosion and Shock Waves*[18]; for more on radiation phenomena, see Modest (2003) and Tessé and Lamet (2011); and for combusting mixtures, see Pollard and Candel (2002).

6.9. Acknowledgments

The author would like to thank Hayat Prud'homme for the patient proofreading of successive versions. The author also extends his gratitude to Joël Sommeria for his expertise and constructive criticism.

18 https://www.researchgate.net/journal/1573-8345_Combustion_Explosion_and_Shock_Waves

6.10. Appendices

The two appendices that follow complement the body of this chapter. Appendix A presents the basics and details of the calculations in the Rankine–Hugoniot theory for combustion waves. Appendix B outlines a history of the science of combustion.

6.10.1. *Appendix A: The Rankine–Hugoniot theory*

In section 6.2, we saw that a **diffusion flame** was located at the contact between an oxidant zone and a fuel zone. Thus, it is in a way a **locus of points** at stoichiometric concentration.

The position of the **premixed flame** seen in section 6.3, on the other hand, cannot be determined *a priori*. The combustion phenomenon produced in a homogeneous premix is a **discontinuity wave** resulting from interactions that involve convection, diffusion of heat and chemical species, and the chemical reaction, with its activation energy and the energy released. The Rankine–Hugoniot theory was developed for this type of discontinuity wave in the stationary case. In order to establish the characteristics of such a wave, we must first establish the **balance equations** for physical quantities at its crossing by treating it as a discontinuity of fluid flow.

6.10.1.1. *Reviewing balance equations*

Let us first recall what we mean by the balance equation of an extensive quantity, F, in a continuous fluid flow. We will then study the balance of F at the level of a discontinuity. As far as possible, at this level of generality, we avoid large mathematical expansions, which may be useful, however, if we wish to move on to applications.

6.10.1.1.1. Continuous balance equations

The balance equation of a quantity F in an arbitrary volume (V) in a **continuous medium** can be formally written as:

Variation of F in *(V)* + **Exit flux** of F across *(∂V)* = **Production** of F in *(V)*

This balance equation can be translated mathematically into the local balance equation of the quantity F in an arbitrary, connected control volume V (Figure 6.19) limited by its boundary *(∂V)*.

If there exists a frame of reference in which the state of the system does not vary over time, the evolution is **stationary** there. By then choosing the fixed border *(∂V)*, the first term of the balance equation disappears.

If the quantity F is conservative, the production is zero and, by definition, the right-hand side of the balance equation disappears.

For **steady evolution**, the non-conservative balance equation has the form

Exit flux of across *(∂V)* = **Production** of F in *(V)*

Mathematically, we write $\int_{(\partial V)} \mathbf{J}_F \cdot \mathbf{n}\, dS = \int_{(V)} \dot{W}_F\, dV$ for any volume *(V)*. It is demonstrated that locally

$$\nabla \cdot \mathbf{J}_F = \dot{W}_F$$

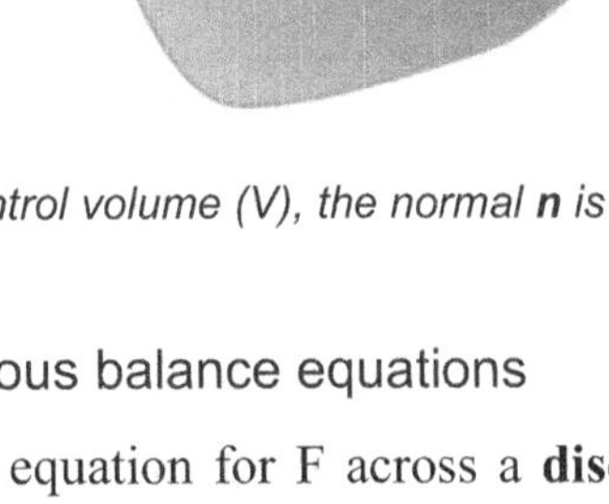

Figure 6.19. *Control volume (V), the normal* ***n*** *is at boundary(∂V)*

6.10.1.1.2. Discontinuous balance equations

The general balance equation for F across a **discontinuity surface** in a flow can be formally written in the same manner as in a continuous medium. However, this time the volume (V) is replaced by a portion of the real interface (S) (Figure 6.20), delimited by a closed curve (C). If the discontinuity is stationary, this means there is a reference frame in which it is at rest and where the first term of the balance equation – the term for the temporal variation of F in (S) – is cancelled.

In the absence of any flux across the line (*C*), which notably means that surface tensions are neglected, the following relation remains:

$$(\mathbf{J}_{F2} - \mathbf{J}_{F2}) \cdot \mathbf{N} = \dot{W}_{aF}$$

which means that the flux jump $\mathbf{J}_F$, projected on the normal **N** as it crosses (*S*), is equal to the production of F **per unit area** (index *a*). We also write $J_{FN2} - J_{FN1} = \dot{W}_{aF}$, with the index N for the projection normal to (*S*).

The jump equation thus obtained is applicable to shock waves in a simple, compressible fluid:

– For conservative quantities: mass, momentum and total energy, we have $\dot{W}_{aF} = 0$ and the balance equation for F is locally written

$$J_{FN2} - J_{FN1} = 0$$

– For non-conservative quantities such as entropy, S, there appears a right-hand side, which is the production rate of the quantity F and within the interface

$$J_{FN2} - J_{FN1} = \dot{W}_{aF}$$

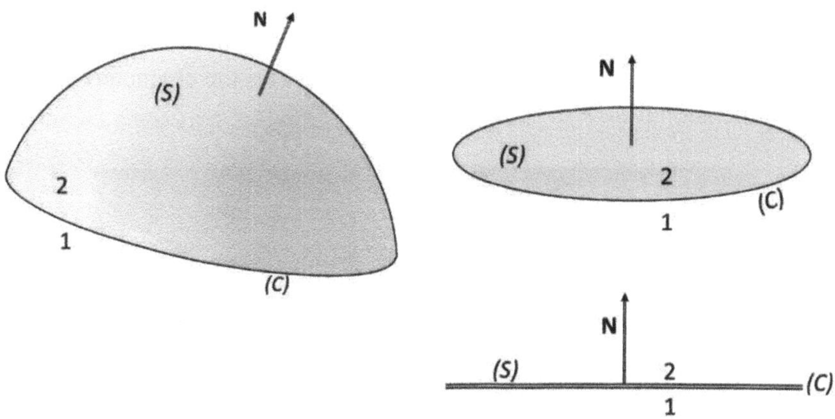

Figure 6.20. *Discontinuity interface. The normal **N** to the discontinuity is oriented in the direction of the relative velocity of the fluid which goes through it. Left: the curved interface; top-right: planar interface; bottom: profile view*

These balance equations are applicable to pure bodies as well as mixtures.

In a non-reactive gas, the only discontinuity waves are compression waves, called **shockwaves,** which have the property of possessing a *supersonic propagation velocity, with respect to the upstream gas.* Expansion waves, such as the centered waves, do exist but are continuous phenomena.

This is not the same thing in a reactive gaseous mixture, as we will see (in section 6.10.1.3) two types of discontinuity waves may develop in the same mixture.

6.10.1.2. *A normal stationary shockwave*

Among the simplest examples of discontinuities, we have normal (or planar) shockwaves (Jaumotte 1971). This problem has already been widely studied. The formulae for normal stationary waves in a perfect gas are well known.

In addition to **normal shockwaves**, there exist **oblique shockwaves** and curved shocks. Figures 6.21 and 6.22 illustrate normal, stationary shockwaves.

Figure 6.5 shows the trajectory of the normal shockwave generated in front of a piston with constant velocity, and indicates the characteristic zones for unidimensional flow in the plane (x, t).

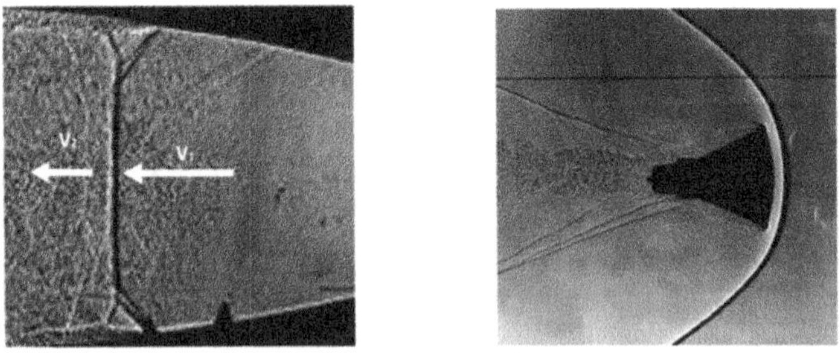

a) b)

Figure 6.21. *Visualizations of shocks. a) Normal shock in the divergent part of a Laval nozzle and b) curved shock in front of a body re-entering the atmosphere, comparable to a normal shock in the vicinity of the axis of symmetry*

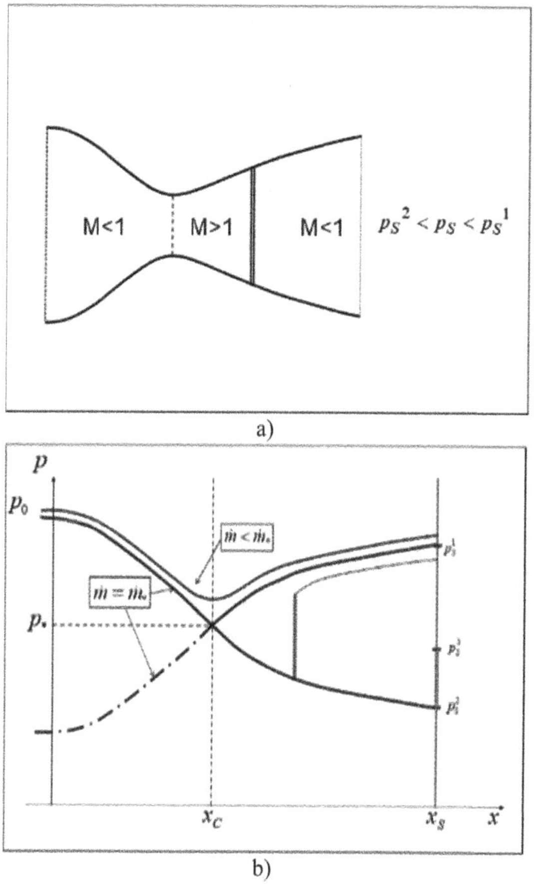

Figure 6.22. *a) Diagram of a Laval nozzle with a shock in the divergent part. b) Solutions for flow in a Laval nozzle. The profile for the nozzle is given by the function A = A(x). The curves that correspond to the critical mass flow $\dot{m} = \dot{m}_*$ are asymptotes for continuous internal flows; the part depicted in dot-dashed lines is fictional*

The balance equations for a normal stationary shockwave are as follows:

$(\rho_2 u_2 - \rho_1 u_1) = (\dot{m}_2 - \dot{m}_1) = 0$ (mass)
$\dot{m}\ (u_2 - u_1) + p_2 - p_1 = 0$ (momentum)
$\dot{m}\ (h_2 - h_1 + u_2^2/2 - u_1^2/2) = 0$ (energy)
$\dot{m}\ (s_2 - s_1) \geq 0$ (entropy)

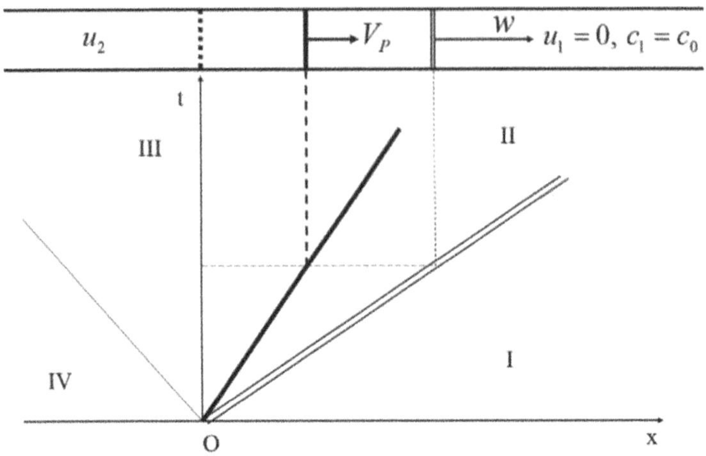

Figure 6.23. *Motion on either side of the piston with uniform velocity V_P. The trajectory of the piston is shown by thick lines and that of the shockwave with velocity w is depicted in double lines. Ahead of the shock, in region I, the gas is at rest $u_1=0$; the speed of sound is c_0*

In these equations, $\dot{m}$ is the unit mass flow (i.e. per unit area); ρ_i is the density I = 1, 2; p_i is the pressure; h_i is the enthalpy per unit mass; u_i is the relative velocity with respect to the wave; and s_i is the entropy per unit mass.

We set $p = p_2/p_1$, $\vartheta = \rho_1/\rho_2$, $\gamma = c_p/c_v$, $M_1 = u_1/c_1$ is the Mach number for the upstream flow, $c_1 = \gamma\ p_1/\rho_1$ is the speed of sound, c_p and c_v are the specific heats per unit mass of the gas.

The flow rate line has the equation: $(p - 1) / (\vartheta - 1) = -\gamma\ M_1^2$, and the production rate of enthalpy $\dot{W}_{Sa}$ is always positive or zero, according **to the second law of thermodynamics**. Considering the ideal gas laws of state: $\dot{m}(s_2 - s_1) = \dot{m}\ c_v\ ln\ (p\ \vartheta^{\gamma})$, with p and ϑ denoting the pressure and specific volume as ratios of their upstream values, respectively.

Figure 6.24(a) depicts the graphical determination of the functional point of the normal shockwave in the Clapeyron plane and Figure 6.24(b) shows the entropy jump along the Hugoniot adiabatic.

A shockwave corresponds to the thin flow zone, thinner than 1 mm, which is the **transition between the upstream zone where the velocity is supersonic and the downstream zone where it is subsonic**. As the shock passes through, there is a strong increase in pressure. Thus, during their crossing of the shockwave, the fluid particles undergo an extremely strong deceleration, which may be as high as 10^9–10^{10} m /s^2, or around a billion *g*. This violent impact, which produces a real shock on the particles, is at the origin of the name given to this transition zone (Lallemand 2014). The shockwave is always a compression discontinuity.

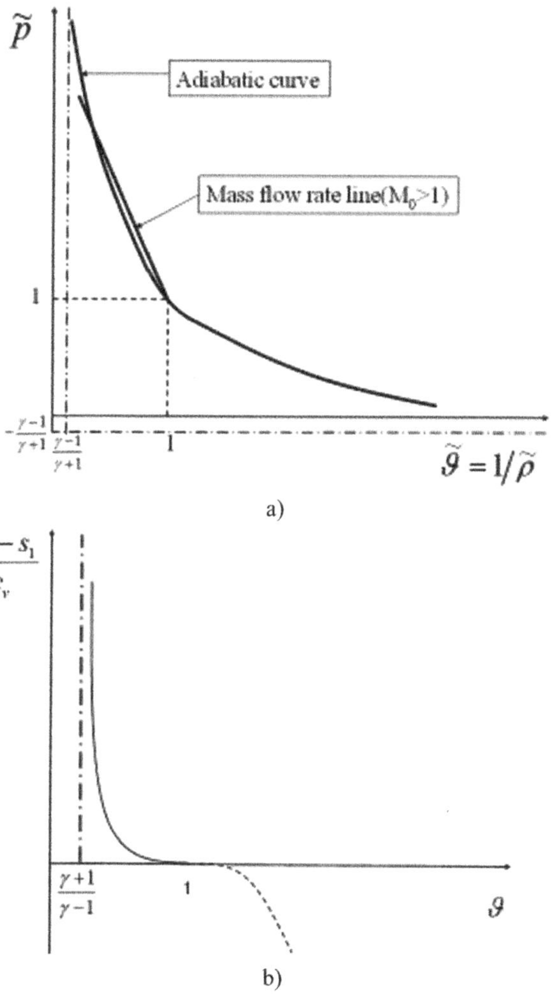

Figure 6.24. *The Hugoniot adiabatic in the planes a) ϑ, p and b) ϑ, s. The Mach number for the upstream flow with respect to the wave is: $M_1 = u_1/c_1$. A shock only occurs for $M_1 > 1$*

6.10.1.3. *Stationary combustion waves*

6.10.1.3.1. Diagrams

The Rankine–Hugoniot equations characterize a flat discontinuity wave in the presence of a **chemical combustion reaction**. The basic equations are similar to the shockwave equations. Thus, the discontinuity equations are written, consecutively for *mass, momentum and energy*, as for the shockwave:

$(\rho_2 u_2 - \rho_1 u_1) = (\dot{m}_2 - \dot{m}_1) = 0$ (mass), $\dot{m}\ (u_2 - u_1) + p_2 - p_1 = 0$ (momentum),

$\dot{m}\ (h_2 - h_1 + u_2^2/2 - u_1^2/2) = 0$ (energy),

however, to this we must add the equations of state for the mixture, which are partly different from those of the shockwave.

By simplifying and taking an average specific heat c_p, we obtain:

$$h_2 - h_1 = -\Delta h + c_p\,(T_2 - T_1)$$

whereas when the shock passes without combustion, we have: $h_2 - h_1 = c_p\,(T_2 - T_1)$.

By eliminating u_1 and u_2 between the equation for mass and the equation for momentum and by writing $\vartheta = \rho_1/\rho_2$, $p = p_2/p_1$, we obtain, after simplification, the following two equations called the "*Rankine–Hugoniot relations*", where $M_1 = u_1/c_1$, the ratio of the velocity to the speed of sound in fresh gases, is the upstream Mach number

$(p - 1)/(\vartheta - 1) = -\gamma\, M_1^2$ (flow rate line)
$p = [(1 - \gamma)\,\vartheta + 1 + \gamma + 2\gamma\,\tau]\,/[(1 + \gamma)\,\vartheta + 1 - \gamma]$ (Hugoniot adiabatic)

that describe, respectively, the normal of the flow rates and the Hugoniot adiabatic curve. Having the upstream Mach number M_1 makes it possible to solve the system. In Figure 6.25, the arc $F_1\,F\,F_2$ of the Hugoniot adiabatic is in the zone of positive entropy variation and makes it possible to have an expansion discontinuity wave. The shock, represented by a thin line, does not present such a wave.

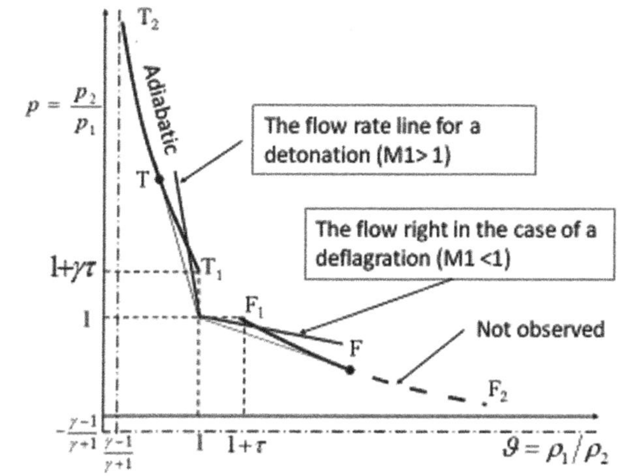

a)

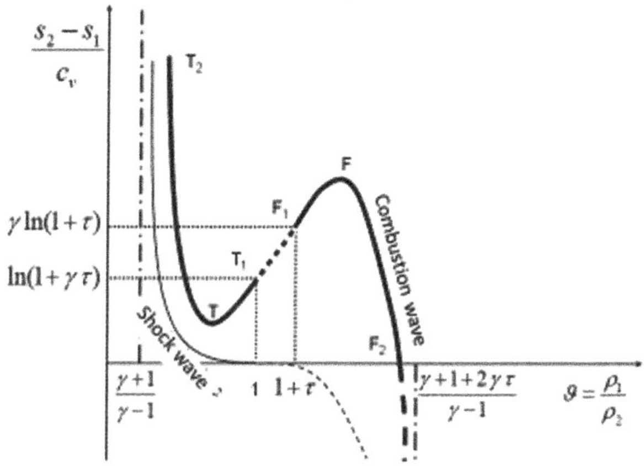

b)

Figure 6.25. *a) The flow rate line (in both cases) $M_1 = u_1/c_1$ denoting the Mach number for the wave with respect to the fresh gases, and the Hugoniot adiabatic. The functional points are found at the intersection of these curves. b) Production rate of entropy when the combustion wave passes along the Hugoniot adiabatic. Notations: isentropic coefficient $\gamma = c_p/c_v$, entropy jump $(s_2 - s_1)$, density ρ, reduced energy released $\tau = \Delta h/c_pT_0$*

6.10.1.3.2. The results

In Figure 6.25(a), the functional points are located at the intersection of the flow rate line at the given M_1 and the Hugoniot adiabatic; however, Figure 6.25(b) indicates the conditions for the existence of solutions, as the entropy variation (s_2 - s_1) must be positive. The presence of a minimum on the entropy variation curve can also be noted. This minimum corresponds to a stable point in the Clapeyron plane in Figure 6.25(a).

This simplified theory (Barrère and Fabri 1971), summarized in section 6.3.1, takes into account the different **stationary planar combustion waves** in a *laminar regime* that can occur in a rectilinear tube. This tube is initially filled with a mixture of oxidant and fuel and its walls are *adiabatic* (i.e. they do not allow heat exchange with the exterior). Igniting one end of this tube first of all allows a slow wave to progress through the fresh gases. The velocity of this wave is of the order of m/s and it is called a **deflagration** wave, being nothing but a premixed planar flame. Then, over time, this wave accelerates and becomes a very rapid **detonation** wave, progressing at supersonic speed.

To summarize: the **compression wave** exists as in the non-reactive case; it is a shockwave boosted by the energy of the chemical reaction and, is thus more intense and totally supersonic (2000–3000 m/s), *called a detonation wave*. However, the existence of **discontinuous expansion waves** is also made possible by the laws of thermodynamics. These are *deflagration waves*, which are clearly subsonic.

In a system of axes associated with the wave, the phenomenon may be schematized as in Figure 6.26:

– upstream of the wave ($x < 0$), the mixture of fresh gas moves at a constant velocity $\mathbf{v}_1$;

– as it passes through ($x = 0$), the fluid is subject to a **discontinuity**;

– downstream of the wave ($x > 0$), the mixture of burnt gases moves with a uniform motion at the velocity $\mathbf{v}_2$.

In the zone with velocity $\mathbf{v}_1$, the concentrations are those of the fresh gases; the temperature is too low for chemical reactions to occur. The mixture is frozen. In the zone with velocity $\mathbf{v}_2$, we have a gaseous mixture in chemical equilibrium. The enthalpy h_2 therefore relates to a gas of a different nature from gas (1). The difference in enthalpies corresponds to the energy released by the reaction to which is added the enthalpy of the heating of the gas between T_1 and T_2.

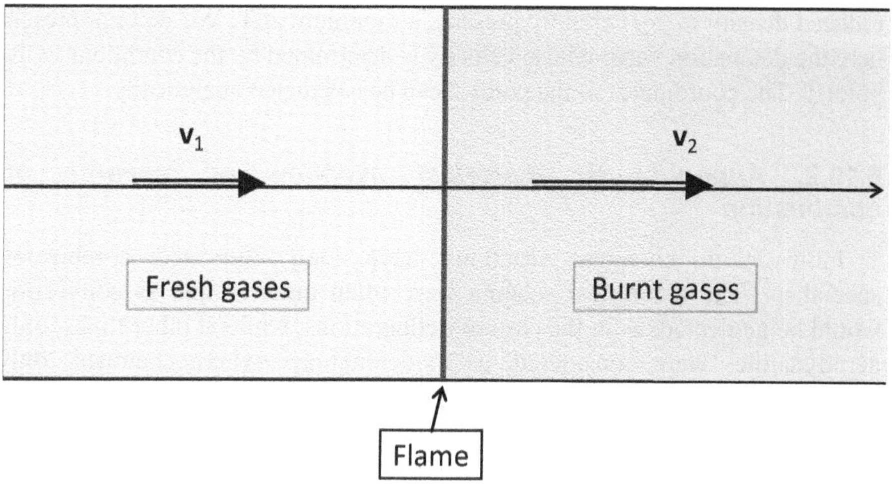

Figure 6.26. *Rankine–Hugoniot configuration: representation of the steady motion at the scale of the dimensions of the tube in a system associated with the wave; straight adiabatic tube initially filled with a mixture of oxidant and fuel; v_1 is the laminar adiabatic combustion rate. Here, the activation energy of the reaction is very high, which explains why the flame is thin. For a color version of this figure, see www.iste.co.uk/prudhomme/fluid1.zip*

In this simplified theory, the solution points may thus be obtained *graphically* by the intersection of a curve and a straight line in the volume–pressure plane. Figure 6.25(a) depicts both cases: deflagration for $M_1 < 1$ and detonation for $M_1>1$, the operational point is found at the intersection of the flow rate line and the Hugoniot adiabatic, which has a hyperbolic shape.

The deflagration wave with velocity $u_1 = s_L^0$ ($_L$ for laminar and ^{0} for adiabatic) corresponds to a point on the branch F_1F where the variation in pressure is low and where an increase in the specific volume indicates high dilation. However, it is not possible to immediately find the exact position of this point. *This was the subject of section 6.4.2.*

The point of detonation on the branch T_1T_2 is easier to determine. The detonation wave is produced at the point T, called the Chapman–Jouguet detonation point. It is possible to demonstrate that this point T, where the flow rate line is a tangent to the Hugoniot curve, is the only stable point on this branch. For this, it is sufficient to analytically calculate the variation in reduced entropy $(s_2 - s_1)/c_v$ at the crossing of the wave. Its evolution, represented in Figure 6.7(b) along the Hugoniot adiabatic as a function of the reduced density ρ_1/ρ_2, certainly presents a minimum at T. We will not present here the detonation wave whose velocity is determined by the conditions at the point T. The coordinates of the point T can be calculated analytically.

6.10.2. *Appendix B: Historical overview of research on combustion*

Flows with chemical reactions have long interested combustion specialists. The theoretical schema were often quite simple as convection would be neglected, as in the case of deflagrations; while at other times, only aerodynamics were considered, as in detonations, where chemistry only came in due to the contribution of the combustion reaction energy.

However, in many problems, the situation has become more complex and disciplines as varied as aerodynamics, thermodynamics and chemistry are simultaneously involved and intertwined. This is the case with combustion in industrial furnaces or propulsion, with problems related to pollution, fire, the penetration of bodies re-entering planetary atmospheres at very high velocities and the problems with reactive boundary layers. Similar situations are found in metallurgy and in certain industrial reactors.

With research work being led by scientists from many disciplines, the term *Aerothermochemistry* has emerged, proposed by Theodore von Kármán, to designate this new science. Combustion has supported the evolution of knowledge, based on the work of eminent precursors such as Duhem, Jouguet, Hirschfelder, Prigogine, De Groot, Napolitano and Zeldovich. Several important texts have been published. In the case of combustion, we can cite the book by Williams (1985a, b) and those of Barrère (Barrère and Prud'homme 1973), and the works of Peters (2000) and Poinsot (Poinsot and Veynante 2005) which have updated content on aspects related to turbulence.

The Combustion Institute is a non-profit scholarly society, founded in 1954 by Bernard Lewis and Hoy Hottel, which aims to promote studies in the field of combustion. The current president of the Combustion Institute is Philippe Dagaut[19] who succeeds James F. Driscoll for the period 2020–2024. The Institute is based in Pittsburgh in the United States and has around 35

19 Philippe Dagaut, President, The Combustion Institute, Research Director at CNRS, Director of ICARE Orleans, France, and Director of the Excellence Center Labex CAPRYSSES, France (see https://www.combustioninstitute.org/about-the-institute/presidents-welcome/).

branches in various countries. France is represented by the French Combustion Group (*Groupement Français de Combustion*). The institute organizes an annual symposium whose works are published and where various awards are presented.

Several journals are associated with the Combustion Institute, including: *Combustion and Flame, Proceedings of the Combustion Institute, Progress in Energy and Combustion Science, Combustion Science and Technology, and Combustion Theory and Modelling.*

Let us cite three scientists who are no longer with us and who have made significant contributions to the field of combustion:

– I.B. Zeldovich (1914–1987)[20] was a Soviet astrophysicist and theoretical physicist. He developed the theory of chain reactions. He made important contributions in the fields of adsorption and catalysis, shockwaves, nuclear physics, particle physics, astrophysics, cosmology and relativity.

– M. Barrère (1920–1996), researcher. He was the Scientific Director of Energetics at the National Office for Aerospace Studies and Research (ONERA). He published several books, including *Rocket Propulsion* and *Fundamental Equations of Aerothermochemistry* (Barrère and Prud'homme 1973; Barrère 1957).

– N. Peters (1942–2015) was professor at the RWTH Aachen University in Germany and was one of the leading world authorities in the field of combustion engineering. He headed the *Institut für Technische Verbrennung* (Institute for Combustion Technology). He published the reference volume *Turbulent Combustion* (Peters 2000).

While not providing an exhaustive list, we must also add the following names:

– B. Lewis (1899–1993), already cited with respect to the creation of the Combustion Institute, was an American chemist of British origin and a major figure in the field of combustion[21]. From 1946 onwards, he was also the

20 It is thanks to him that we have the Zel'dovich number $Ze = E_a(T_b - T_u)/RT_b^2$ representing the reduced activation energy. E_a is the molar activation energy of the chemical, T_b and T_u are the temperatures of the burnt gases and fresh gases, respectively, and R is the ideal molar constant. See Zeldovich (1992).

21 The Lewis number is commonly used to compare the effects of thermal conduction with the effects of the diffusion of chemical species: Le = k/D.

Head of the Explosives and Physical Sciences division of the United States Bureau of Mines in Pittsburgh.

– Adolphe Van Tiggelen (1914–1969), editor of a two-volume book *Oxydations et Combustions* (Oxidations and Combustions) (Van Tiggelen *et al.* 1968), was a Belgian researcher and professor at the Catholic University of Louvain, UCL. He made significant contributions to the understanding of flame processes. In 1961, he was awarded the Francqui Prize on Exact Sciences.

– Luigi Gerardo Napolitano (1928–1991), director of the Aerodynamics Institute under the University of Naples "Federico II", and also taught at Berkeley, the Sorbonne and Poitiers. His field of research was especially vast and included the thermodynamic modeling of multi-reactive gaseous mixtures and the Marangoni effect in microgravity. He was the President of the International Astronautical Federation. He was also the founder and first president of the Microgravity Advanced Research and Support Center (MARS) and of the Italian Centre for Aerospace Research, CIRA.

We will not provide an updated list of modern-day combustion scientists here. The reader can find these in the headings of numerous articles available in journals, whether or not these have been cited in this appendix. These are also often works published by teams rather than individuals. Some texts are given in the reference section that follows (Borghi *et al.* 1985; Borghi and Champion 2000; Pollard and Candel 2002, Chapter 6). We must also mention the book by Jean-Charles Herpin (1824), dating back over two centuries, which may be entertaining and instructional for readers today to get an idea of chemical experimentation during this period. While not a specialist, Herpin selected experimental curiosities, some of which are related to combustion.

6.11. References

Aldushin, A.P., Malomed, B.A., Zel'dovich, Y.B. (1981). Phenomenological theory of spin combustion. *Combustion and Flame*, 42, 1–6.

Barrère, M. (1957). *La propulsion par fusées*. Dunod, Paris.

Barrère, M. and Fabri, J. (1971). Ondes de choc avec combustion. In *Chocs et ondes de choc*, Jaumotte, A.L. (ed.). Masson, Paris.

Barrère, M. and Prud'homme, R. (1973). *Equations fondamentales de l'aérothermochimie*. Masson, Paris.

Bharath, N.T., Rashkovskiy, S.A., Tewari, S.P., Gundawar, M.K. (2013). Dynamical and statistical behaviour of discrete combustion waves: A theoretical and numerical study. *Physical Review E*, 87, 042804.

Borghi, R. and Champion, M. (2000). *Modélisation et théorie des flammes*. Editions Technip, Paris.

Borghi, R., Clavin, P., Liñán, A., Pelcé, P., Shivashinsky, G.I. (1985). *Modélisation des phénomènes de combustion*. Eyrolles, Paris.

Bray, K.N.C. (1980). Turbulent flows with premixed reactants. In *Turbulent Reacting Flows*, Libby, P.A. and Williams, F.A. (eds). Springer-Verlag, Berlin.

Bray, K.N.C. and Peters, N. (1994). Laminar flamelets in turbulent flames. In *Turbulent Reacting Flows*, Libby, P.A. and Williams, F.

Burke, S.P. and Schumann, T.E.W. (1928). Diffusion flames. *First Symposium on Combustion*, Swampscott, MA, 10–14 September.

Damköhler, G. (1940). Der Einfluss der Turbulenz auf die Flammengeschwindigkeit in Gasgemischen. *Zeitschrift für Elektrochemie und angewandte physikalische Chemie*, 46(11), 601–652.

Dorey, L.-H. (2012). Modélisation des phénomènes couplés combustion – formation des suies – transferts radiatifs dans les chambres de combustion de turbine à gaz. Thesis, ECP-ONERA, Paris.

Emmons, H.W. (1956). The film combustion of liquid fuel. *Zeitschrift für angewandte Mathematik und Mechanik (ZAMM)*, 36, 60–71.

Glorian, J. (2014). Cinétique hétérogène pour la combustion de l'aluminium. PhD Thesis, ENSTA ParisTech, France [Online]. Available at: https://pastel.archives-ouvertes.fr/tel-01144079.

Hebert, D., Coppale, A., Talbaut, M. (2013). 2D soot concentration and burning rate of a vertical PMMA slab using Laser-Induced Incandescence. *Proceedings of the Combustion Institute*, 34, 2575–2582.

Herpin, J.-C. (1824). Récréations chimiques ou recueil d'expériences curieuses et instructives auxquelles on a joint un précis élémentaire de chimie [Online]. Available at: https://books.google.fr/books?id=E0F7zCAKzCoC.

Holt, J.B., Kingman, D., and Bianchini, G. (1985). Kinetics of the combustion synthesis of TiB2. *Mater. Sci. Eng.*, 71, 321–327.

Inoue, C., Izato, Y., Miyake, A., Villermaux, E. (2017). Direct self-sustained fragmentation cascade of reactive droplets. *Phys. Rev. Lett.*, 118, 074502.

Jaumotte, A.L. (ed.) (1971). *Chocs et ondes de choc*. Masson, Paris.

Kirdyashkin, A.I., Kitler, V.D., Salamatov, V., Maksimov, Y.M. (2007). Capillary hydrodynamic phenomena in gas-free combustion. *Combustion Explosion and Shock Waves*, 43(6), 645–653.

Klimenok, K.L. and Rashkovskiy, S.A. (2015). Discrete model of gas-free spin combustion of powder mixture. *Physical Review E*, 91(1), 012805.

Kuznetsov, V.R. (1982). Effect of turbulence on the formation of large superequilibrium concentration of atoms and free radicals in diffusion flames. *Mehan. Zhidkosti Gasa*, 6, 3–9.

Lallemand, A. (2014). Écoulements des fluides – fluides compressibles en monodimensionnel [Online]. Available at: https://www.techniques-ingenieur.fr/base-documentaire/energies-th4/fluides-et-combustion-42213210/ecoulements-des-fluides-be8165/.

Maksimov, Y.M., Pak, A.T., Lavenchuk, G.B., Malborodenko, Y.S., Merzhanov, A.G. (1978). Spin combustion of gasless systems, Tomsk. *Fizika Gorentya I Vzryva*, 15(4), 156–159.

Modest, M.F. (2003). *Radiative Heat Transfer*, 2nd edition. Academic Press, London.

Nicoli, C., Haldenwang, P., Denet, B. (2006). Flame holding downstream from a co-flow injector. *C. R. Mecanique*, 224, 408–413 [Online]. Available at: http://france.elsevier.com/direct/CRAS2B/.

Peters, N. (1980). Local quenching of diffusion flamelets and non-premixed turbulent combustion. Western States Section of the Combustion Institute, Paper WSS 80-4, Spring Meeting, Irvine, USA.

Peters, N. (2000). *Turbulent Combustion*. Cambridge University Press, Cambridge.

Poinsot, D. and Veynante, T. (2005). *Theoretical and Numerical Combustion*, 2nd edition. Edwards, Philadelphia.

Pollard, A. and Candel, S. (2002). *IUTAM Symposium on Turbulent Mixing and Combustion*. Kluwer Academic Publishers, Dordrecht.

Prud'homme, R. (2010). *Flows of Reactive Fluids*. Springer, Berlin.

Prud'homme, R. (2015). Vaporisation et combustion de gouttes dans les moteurs [Onine]. Available at: https://www.techniques-ingenieur.fr/base-documentaire/mecanique-th7/machines-hydrauliques-et-thermiques-fondamentaux-et-concepts-innovants-42171210/vaporisation-et-combustion-de-gouttes-dans-les-moteurs-bm2521/.

Rashkovskiy, S.A., Gundawar, M.K., Tewari, S.P. (2010). One-dimensional discrete combustion waves in periodical and random systems. *Combust. Sci. and Tech.*, 182(8), 1009–1028.

Tessé, L. and Lamet, J.M. (2011). Radiative transfer modeling developed at ONERA for numerical simulations of reactive flows. *Aerospace Lab Journal*, (2), 1–19.

Van Tiggelen, A., Burger, J., De Soete, G. (1968). *Oxydations et combustions*. Editions Technip, Paris.

Williams, F.A. (1985a). *Combustion Theory*, 1st edition. Addison Wesley Publishing Company, Boston.

Williams, F.A. (1985b). Turbulent combustion. In *The Mathematics of Combustion*, Buckmaster, J. (ed.). SIAM, Philadelphia.

Zeldovich, Y. (1992) *Selected Works of Yakov Borisovich Zeldovich, Volume I: Chemical Physics and Hydrodynamics*. Princeton University Press, Princeton.

List of Authors

Daniel BEYSENS
P.M.M.H., CNRS, ESPCI
PSL Research University
Sorbonne University
Paris
France

Mohamed-Amine CHADIL
MSME Laboratory
Gustave Eiffel University
Paris
France

Christian CHAUVEAU
ICARE
CNRS
Orleans
France

Jean-Luc ESTIVALÈZES
DMPE
ONERA, the French Aerospace Lab
Palaiseau
France

Solène GOUÉNARD
CEA CESTA
Le Barp
France

Mahouton Norbert HOUNKONNOU
University of Abomey-Calavi
Republic of Benin

Stéphane MIMOUNI
EDF R&D
Paris
France

Roger PRUD'HOMME
Institut d'Alembert
Sorbonne University
CNRS
Paris
France

Stéphane VINCENT
University of Paris-Est
Marne-La-Vallée
France

Index

A, B

Aslam extension, 86, 88, 91–116
asymptotic expansions, 24, 31
Brownian motion, 151, 152, 156, 158, 161, 163

C, D

capillarity, 20
colloids, 151, 152, 156–159, 161, 163, 164
combustion, 167
critical point, 123, 129, 131, 142, 151, 152, 154, 155, 159, 163
curvilinear coordinates, 1, 11, 12, 14, 15
drag coefficient, 85, 90, 100, 103

E, F

evaporation, 140, 144
flame, 2, 6, 8, 10, 24, 28–30, 169, 170
  diffusion, 167, 171–174, 176, 179, 181, 186, 187, 189, 198, 201
  premixed, 167, 171, 172, 179, 181, 183, 186–188, 196, 199, 201
fluctuations, 151–155, 159, 161, 163, 164
fluid
  critical, 121, 132, 133, 135, 142
  supercritical, 121–124, 133, 134, 139, 146

G, I

gaseous phase, 122, 140, 193, 195
interfaces, 121, 137, 140, 146
  large, 38, 41–45, 65, 68, 78
irreversibility, 24

L

large-eddy simulation (LES), 37, 39–42, 46, 48, 56, 60–62, 64, 65, 67, 68, 78
liquid mixture, 153, 160, 161, 163

M, N

microgravity, 123, 132, 135, 144, 146
multifield approach, 37, 41–43, 45, 61, 62, 64, 68, 77
Nusselt number, 85, 90, 105–107, 110

P, S, T

particle-resolved simulations, 84, 86, 113
phase changes, 6, 8, 37, 38, 40–43, 68, 70, 75, 77, 78
  liquid–vapor, 121, 141, 145
solid rocket motors, 194
soot, 171, 172, 180, 188, 190, 191
sprays, 167, 197, 198
two-fluid model, 37, 40–50, 67, 68, 70, 77, 78
two-phase flows, 38, 41, 43, 61, 78

Summary of Volume 2

Preface

Roger PRUD'HOMME, Stéphane VINCENT, Christian CHAUVEAU and Mahouton Norbert HOUNKONNOU

Chapter 1. Turbulent Channel Flow to Re_τ= 590 in Discrete Mechanics

Jean-Paul CALTAGIRONE and Stéphane VINCENT

1.1. Introduction
1.2. Discrete mechanics formulation
1.3. Turbulent flow in channel
1.3.1. Analysis of a turbulent flow in a planar channel
1.3.2. Model of the turbulence in discrete mechanics
1.3.3. Application to a turbulent flow in a channel with Re_τ = 590
1.4. Conclusion
1.5. References

Chapter 2. Atomization in an Acceleration Field

Roger PRUD'HOMME

2.1. Introduction
2.1.1. Two classic instabilities
2.1.2. Atomization
2.2. Generation of droplets through vibrations normal to the liquid layer
2.3. Rayleigh–Taylor instability at the crest of an axial wave
2.3.1. Size distribution of the drops

2.4. Recent work
2.5. Conclusion
2.6. References

Chapter 3. Numerical Simulation of Pipes with an Abrupt Contraction Using OpenFOAM

Tarik CHAKKOUR

3.1. Introduction
3.2. Modeling an abrupt contraction in a pipe
3.2.1. Euler equations
3.2.2. Stability of the solver
3.2.3. Introducing the model
3.2.4. Boundary and initial conditions
3.3. Numerical results
3.3.1. Results with the boundary and initial conditions I
3.3.2. Results with the boundary and initial conditions II
3.4. Conclusion and future prospects
3.5. References

Chapter 4. Vaporization of an Equivalent Pastille

Roger PRUD'HOMME and Kwassi ANANI

4.1. Introduction
4.2. Equations for the problem
4.3. Linear analysis of the liquid phase
4.3.1. The function $G(u, Pe_L)$
4.3.2. Solution
4.3.3. The depth to which heat penetrates
4.4. Some results
4.4.1. Thermal perturbations
4.4.2. Response factor
4.5. Conclusion
4.6. References

Chapter 5. Thermal Field of a Continuously-Fed Drop Subjected to HF Perturbations

Roger PRUD'HOMME, Kwassi ANANI and Mahouton Norbert HOUNKONNOU

5.1. Drops in a liquid-propellant rocket engine
5.2. A continuously fed droplet

5.3. Equations of the problem
5.3.1. Equations for the gaseous phase
5.3.2. Equations for the liquid phase
5.4. Linearized equations
5.5. Linearized equations for small harmonic perturbations
5.6. Thermal field in the drop when neglecting internal convection
5.7. Conclusion
5.8. Appendix 1: Coefficients that come into play in linearized equations
5.9. Appendix 2: Solving the thermal equation
5.10. Appendix 3: The case of the equivalent pastille
5.11. Appendix 4: 2D representation for the spherical drop
5.12. References

Chapter 6. Study of the Three-Dimensional and Non-Stationary Flow in a Rotor of the Savonius Wind Turbine

Francis RAVELOSON, Delphin TOMBORAVO and Roger VONY

6.1. Introduction
6.2. Mathematical modeling of the problem
6.2.1. Presentation of a physical model
6.2.2. Simplifying hypotheses
6.3. Numerical resolution
6.3.1. Presentation of meshes
6.3.2. Spatial discretization
6.3.3. Temporal discretization
6.3.4. Stability condition for the scheme
6.3.5. Initial conditions
6.3.6. Boundary conditions
6.4. Validation of the results
6.5. Results and discussion
6.5.1. Influence of the advance parameter
6.5.2. Influence of the angular position of the blades
6.6. Conclusion
6.7. Acknowledgments
6.8. References

Other titles from

iSTE

in

Fluid Mechanics

2020

SAD CHEMLOUL Nord-Eddine
Dimensional Analysis and Similarity in Fluid Mechanics

2019

CALTAGIRONE Jean-Paul
Discrete Mechanics: Concepts and Applications

2018

ABABOU Rachid
Capillary Flows in Heterogeneous and Random Porous Media 1

2017

TARDU Sedat
Wall Turbulence Control

2015

CALTAGIRONE Jean-Paul
Discrete Mechanics

FOHR Jean-Paul
Heat and Moisture Transfer between Human Body and Environment

2014

PRUD'HOMME Roger
Flows and Chemical Reactions in Heterogeneous Mixtures
Flows and Chemical Reactions in an Electromagnetic Field

TARDU Sedat
Transport and Coherent Structures in Wall Turbulence

2013

BORGHI Roland, ANSELMET Fabien
Turbulent Multiphase Flows with Heat and Mass Transfer

DÉLERY Jean
Three-dimensional Separated Flow Topology: Critical Points, Separation Lines and Vortical Structures

PRUD'HOMME Roger
Flows and Chemical Reactions in Homogeneous Mixtures

2012

BOUTIER Alain
Laser Velocimetry in Fluid Mechanics

BOUTIER Alain
Laser Metrology in Fluid Mechanics: Granulometry, Temperature and Concentration Measurements

PRUD'HOMME Roger
Flows and Chemical Reactions

2011

MORY Mathieu
Fluid Mechanics for Chemical Engineering

TARDU Sedat
Statistical Approach to Wall Turbulence

2010

COLIN Stéphane
Microfluidics

DÉLERY Jean
Handbook of Compressible Aerodynamics

GUINOT Vincent
Waves Propagation in Fluids – 2nd edition

SOULI Mhamed, BENSON J. David
Arbitrary Lagrangian Eulerian and Fluid-structure Interaction

2009

FAVRE-MARINET Michel, TARDU Sedat
Convective Heat Transfer

2008

LALAUZE René
Physical Chemistry of Solid-Gas Interfaces

PEUBE Jean-Louis
Fundamentals of Fluid Mechanics and Transport Phenomena

SCHIESTEL Roland
Modeling and Simulation of Turbulent Flows

Printed and bound by CPI Group (UK) Ltd, Croydon, CR0 4YY
13/09/2026
14961038-0001